INTERNATIONAL
1488

SNOOPY AND THE SPY

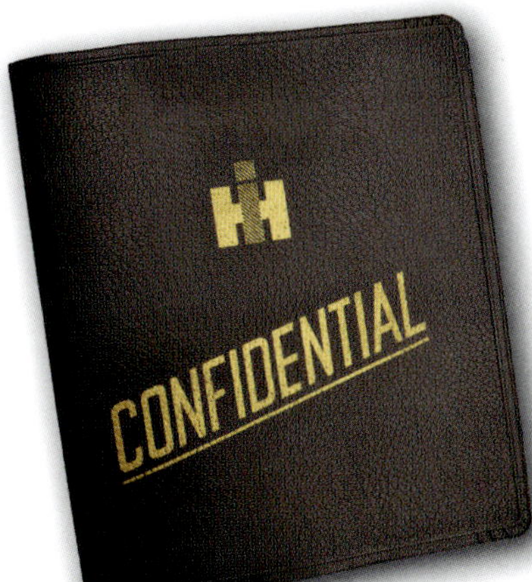

HJ 2

SNOOPY AND THE SPY

A Saga of Espionage, Ingenuity, and the Epic Battle to Save International Harvester

LEE KLANCHER

Octane Press, Edition 1.0, October 2024

On the IH cover: The illustration is a concept drawing of a late model 2+2, a 7488. It has been modified to include the stripe from the 3388 / 3588. *Gregg Montgomery*
On the John Deere cover: This prototype Deere is a wooden model from the mid-1970s created by the Henry Dreyfuss Agency. Note the image is reversed to face the correct direction on the jacket. *Henry Dreyfuss Collection, Cooper Hewitt, Smithsonian Design Museum*
On the frontispiece: A confidential report created for IH—this was a proposal from Hough to create the 4300 for Harvester. *Wisconsin Historical Society*
On the verso page: This is the original unedited drawing of the 7488. The cover drawing is the same, just with a modified hood decal. *Gregg Montgomery*

Hardcover ISBN: 978-1-64234-185-0
ePub ISBN: 978-1-64234-186-7

LCCN: 2024941094

Cover and Interior Designed by Linsey Dodaro
Copyedited by Faith Garcia
Proofread by Christine Florie

OCTANE PRESS

octanepress.com
Octane Press is based in Austin, Texas

Printed in China

In memory of John "Bud" William Youle

(1937-2020)

CONTENTS

Prologue

BEHIND ENEMY LINES

July 1982

New Orleans, Louisiana

When Bud Youle left his hotel room in New Orleans, the early morning air was already a steamy seventy-five degrees, with a dew point that started off oppressive and quickly turned miserable as the mercury rose above ninety.

The hot air of summer in Louisiana infamously creates tropical storms of epic proportions, capable of downing jets and wiping out entire precincts. The storm on this day was not a widow-maker, but it would not be without causalities.

The year 1982 was tough on farmers, equipment makers, and pretty much everyone else in America. This 1976 International tractor is at work near Macon, Georgia. *Author's Collection*

Youle worked in the farm equipment industry, a business that was in the midst of the superstorm of the century. He was a salesman for one of the world's largest tractor manufacturers, International Harvester. He and his company were battling forces just as oppressive as the Louisiana air. Bankers, courts, and the opposition smelled blood in the water. The sharks were circling.

While equipment manufacturers were rolling out new iron, farmers trying to survive the times opted to run old machinery. This beat-up 1960s model is at work tilling for Mr. Sims in Tillay, Arizona, on May 6, 1978. *NARA*

A fiery orator and a force of nature, Youle had spearheaded the sale of a new machine that helped keep Harvester afloat into the 1980s. The previous decade had given the company fresh leadership, but it was short on cash and time. The idea had emerged to create an innovative four-wheel-drive tractor by mating two rear ends from its aging tractor product line, hinging the tractor in the middle. The machines that resulted—the 2+2s—gave farmers a four-wheel-drive machine that was more nimble and better suited to cultivation than the four-wheel drives of the time.

The new models had given Harvester a much-needed sales boost in the late 1970s, and Youle had been the lead disciple in the church of 2+2 tractors. The odd-looking models 3388 and 3588—dubbed "Snoopy" by the competition due to their long, protruding noses—had sold like hotcakes and played a role in driving Harvester to its most profitable year in recent history.

In 1982, Harvester's primary competitor, John Deere, had answers to the Snoopy problem. Harvester had been the dominant tractor maker in the first part of the twentieth century; Deere had taken over the industry lead in 1963. As Harvester pulled out the stops to develop the new machinery needed to regain its lead, Deere outspent it on research and development and was always one step ahead. Every time Harvester rolled out a new innovative machine, Deere answered the call to battle with an improved version of its own.

In the heavy equipment world, creating new machines requires massive amounts of time and money. An all-new machine can require a decade of development and millions—or billions—of dollars. The business is also cyclical, so if a company introduces a machine into a down market, it can take decades or longer to recoup its financial investment.

"American farmers . . . have had two companies pour millions and billions of dollars in the product so that they can farm faster, cheaper, and better than anybody in the world."
– Bud Youle

The fierce competition between Deere and Harvester in the 1960s and 1970s had transformed the agriculture sector, with the two controlling the lion's share of the market and the other brands—Ford, J. I. Case, Allis-Chalmers, Steiger, and White—scrapping for single-digit portions of what was left.

That competition drove farm technology at a rapid rate. "American farmers are blessed," Youle said. "They have had two companies pour millions and billions of dollars in the product so that they can farm faster, cheaper, and better than anybody in the world."[1]

For 1982, the farmer was treated to another round of high-technology mana produced by this epic battle. Deere was introducing a new line of high-horsepower tractors that featured a greatly improved mechanical front drive (MFD) system. The system provides power to the front wheels, but the technology took a long time to get right on tractors, and early versions steered poorly and required a country mile to turn around. The Deere MFD machines—dubbed the 50 Series—purported to have solved the

While Deere and Harvester spent money in a heated technology race, the other makers (like Ford) lost market share. For farmers, the worst farm economy in modern history had them struggling just to survive. This is farmer and farm advocate Edna Harris on her farm in Iredell County, North Carolina, in 1987. *Rob Amberg*

FORD

problems inherent with the current MFD systems. More significantly, these new models were a challenge to Youle's beloved 2+2.

Deere was showcasing the new machines to its dealers in New Orleans. These meetings were standard industry practice for heavy equipment manufacturers, and they were often particularly elaborate when new models are introduced. They were also invite-only events, created specifically for select members of the brand's dealership network.

In this case, Deere had rented the Superdome and planned a grand rollout to show off the new technology, and in term coerce their dealer network to order vast numbers of the new machines. Deere was planning to make the meeting in New Orleans its biggest dealer rollout since 1960, the year it had unveiled the line of machines that would ultimately topple Harvester from the leadership slot in the tractor industry.

Youle had gotten wind of the rollout—which is not surprising, as the two makers tracked each other's movements carefully and generally knew what was coming—and he convinced management that he should be in New Orleans to take measure of the new machines.

"I need to go to New Orleans," Youle told them, "We got to lay out a program to get ready for these people."

He wanted to find out how much horsepower the tractors had and if the technology used to steer the new mechanical front-wheel drive was effective. More broadly, what he was seeking was the inevitable flaws the machines carried so that Harvester could craft its promotional campaigns to attack those weak points.

With bankers, courts, and competitors comprising a herd of hungry dragons breathing down the neck of Harvester in 1982, the stakes for maintaining market share had never been higher for the red faithful.

Youle remembered the response to his travel proposal being quick and positive. "Okay, you go. Take off. You go down to New Orleans."

Spying on the opposition was part of the game in tractors ever since an Iowa farmer spotted a strange rig in South Dakota. He took the idea home and built his own . . . in the 1880s. The games got rougher as the competition for sales grew fiercer in the 1910s and 1920s, with field demonstrations regularly including sabotage, lies, and slander—on occasion dropping the pretense of civility and devolving into bouts of fisticuffs.

By the time Youle strolled onto the grounds of the Superdome, tractor maker wars were waged gathering intelligence. Youle made his first foray into enemy territory clad in nondescript clothing. While most of corporate America lives and dies in drab attire, this is not the modus operandi of denizens of the tractor world.

Harvester spy Bud Youle stayed in a New Orleans hotel during his secret mission, likely much like this one photographed in 1982. *John Margolies, Library of Congress, LC-MA05- 8432*

The agricultural equipment world is defined by color. Machines are painted a distinctive branded color, a bright tone that can be spotted from a mile away. The colors in 1982 ranged from primary and vivid—Ford in blue, John Deere in green, and Harvester in red—to eye-watering and fluorescent—the small and agile Kubotas from overseas in high-viz orange and the high-horsepower Steigers from North Dakota (the Cadillacs of tractors) sporting a limey bright green with cartoony predator names such as "Puma" and "Panther."

The color coats much more than just the machines. Branded gear abounds. Employees are covered in heavily branded work wear emblazoned in their corresponding company color. Dealership walls and signs are bright and bold, and advertisements feature heavy dollops of the appropriate shade.

What is perhaps most unique about this is the depth of the color loyalty found in tractor owners. Most farms are defined by brand loyalty, easily identified simply by color. If you run Harvester equipment, you run red. The self-anointed members of this clan spend a good portion of their lives clad in red clothing, often T-shirts emblazoned with "If it isn't red, leave it in the shed" or the like. Barns, silos, houses, and pickup trucks often wear the same color, and tractor brand-colored weddings are fairly common.

Color defines tractors, and tractors define farmers.

For Youle to drop his colorful Harvester brand wear, well, that's a stealth move right there. He walked around the outside of the Superdome and saw green machines being prepped to go inside for the show. He asked around and learned a few things. Eyeballed the new front-wheel-drive system. Chatted up the green guys for horsepower ratings.

New Orleans jazz man.
Carol Highsmith, Library of Congress, LC-HS503-1626

He had hunches about what was missing, about what didn't look right, but he didn't know the score.

He wanted to go inside, wanted to hear what they had to say. Get the scoop on the marketing. Get a jump on the competition.

If he got caught, he would set off an "international incident." Sneaking around in parking lots was one thing—but faking your way into an event that was for green guys only? That was high treason. Game over. Fireworks.

If he got caught, the blowback would be off the charts.

If he got caught, the blowback would be off the charts.

Then he had an idea: He had been in the Superdome before, and he knew there was a bar inside, way up top, that had a view of the floor. If a man could get up there, he thought, he could hear the announcements. See what was going on. Sit quiet-like, sip a few longnecks, and slip out the back at the end of the day.

Easy-peasy.

Treason be damned. Times were hard, and Youle was not one to stay on the sidelines. He crossed the superheated pavement of the Superdome grounds, suffering in the swampy July air. He strode inside the air-conditioned oasis of enemy territory, a red spy in a sea of top-secret technology intended only for the eyes of the green faithful.

His actions that day—much like the thick, wet air of New Orleans—would create a storm of epic proportions that would reach to the top echelons of the world's two largest tractor makers, jeopardize Youle's career, and shake up an embattled giant.

This is that story.

Home Newspaper of the Calumet Region
Dead Baby Show Closed
Steel Firm Sold
OFFICIAL NCAA BASKETBALL BOX SCORE
If you're shopping the 40- to 80-hp tractor class...
Here are the questions you should have answered
INTERNATIONAL AXIAL-FLOW COMBINE
DIAMOND
MR. WRESTL
All Ringside $4.50

Chapter One

BATTLE ROYALE

1970s

AGAINST ALL ODDS

September 1977

Washington, Iowa

In late September 1977, Maurice and Dennis Crawford's expansive eight hundred-head beef farm near Washington, Iowa, would transform into a bit of a circus. The farm would host 434 exhibitors jammed into tents and a portable sixty-five-foot-tall dome, a horde of corporate sorts in brand colors as well as independent vendors hawking wares that ranged from nematodes (invisible worms that help things grow) to automatic hog feeders.

Local church groups gathered to feed the estimated crowd of 250,000 expected to walk the grounds and eyeball the latest farm equipment and technology. Mrs. Richard Brinning of the United Methodist Church in Washington, Iowa, started preparing food early in the morning, working in the unseasonably cool forty-degree air to ready her and her crew to feed eight hundred to one thousand people that day—and raise good money for the church.

The event was the annual Farm Progress Show, the country's premier showcase for new agricultural equipment. In those days, the show was held on a farm somewhere in the heartland. The hosts were rewarded for their hospitality with compensation for the crops they weren't able to harvest and with deals on equipment from the vendors. Sometimes, the host got a new building if one of the manufacturers wants to showcase its latest silo or outbuilding.

Prior to 2005, the Farm Progress Shows were held on private farms in the Midwest. In 1977, the show was held on the Crawford farm near Washington, Iowa. *NARA / USDA*

The 1977 show on the Crawford Farm was

dampened by a depressed farm economy and widespread drought in the western part of the country. This didn't stop the manufacturer's from rolling out new hardware. Each of the major agricultural implement companies had some new machine in the lineup.

This is the Versatile Model 1080, a thirty-foot-long beast with a Cummins 600-horsepower engine, eight-wheel drive, and a closed-circuit television that functioned as a rearview camera. *Peter Simpson Collection*

The horsepower race was in full swing, with large four-wheel drives all the rage. The showstoppers of the era were high-horsepower behemoths, most of them one-offs built by the makers in an escalating war to create the biggest beast on the field.

Steiger Tractor out of Fargo, North Dakota, had its resident genius Paul Nystuen meld two of its big Panthers together to create the 650-horsepower Panther Twin.[2]

Versatile out of Canada had built an insane eight-wheeled creation with a 600-horsepower engine out back, including gee-whiz technology such as a nine-inch television screen built into the dash.

The king of the fields, however, was the 760-horsepower Big Bud 747 built in Havre, Montana, by Ron Harmon. The monstrous machine measured twenty-seven feet long, was twenty feet tall, and had a one thousand-gallon fuel tank. Crowds would gather then and now to see it at work.

Among all the giant tractor shenanigans, the two industry leaders—top dog John Deere and the old giant Harvester in second place—had to find a way to stand out in the crowd. Harvesting crops in field demonstrations was a time-honored way to sell equipment, and each year the harvesting equipment of all makers took to those Farm Progress demonstration fields to do their darndest to show up the competition and gain customers.

For 1977, some seven hundred acres of the Crawford's ground was planted with corn and soybeans.[3] The battle for crop supremacy was particularly heated that year. Harvester had an entire lineup of new models—the 86 Series. While this gave it something to showcase at the Farm Progress Show, the fact was the tractors had a nice new paint scheme and some modest styling changes . . . and not much else.

A part of the Harvester team's gambit to win back the lead in the agricultural equipment industry was a new combine introduced in 1977, the Axial-Flow. The machine used rotary threshing, a technology which was very time-consuming to develop. Harvester was willing to expend the time and energy to finish the tech for two key reasons. First, its combine engineering team was bullheaded and independent—not to mention located 170 miles away from corporate headquarters. The combine crew had a soft spot for the tech and had been developing it off and on since the 1950s (!), more or less off the books.

Second, Harvester was flat-out desperate. Harvester had been the industry giant since it formed in 1902, and Deere had been content with the second position for decades. The winds began to shift in the 1950s when John Deere locked a small group of its best engineers away in an old butcher shop with blacked-out windows and told them they had a clean slate to build a machine that would put Deere out front. The New Generation tractors were introduced in 1960, and they were indeed a leap forward in terms of power, features, style, and comfort compared to any other large tractor on the market. By 1963, Deere's sales were the highest in the ag equipment business, and it would not look back.

Harvester did its best to fight back, but it was fighting from behind due to two decades of insufficient research and development coupled with some expensive missteps that included the heavy expansion of its construction equipment line—a market in which Caterpillar had a stranglehold—and an oddball (and ultimately doomed) venture into home appliances.

"IH was continually in the shadow of competition, primarily Deere,"

> "IH was continually in the shadow of competition, primarily Deere. . . . We needed something spectacular."
> –Engineer Don Murray

Giant high-horsepower one-off machines were all the rage in the late 1970s. This example is a Big Bud 525/50, one of the most powerful tractors built in Montana. Note that this is a slightly later model that debuted in 1979—the KT models would have been at the 1977 Farm Progress Show. *Marcus Pasveer*

wrote Harvester's chief engineer for the combine program, Don Murray. "We needed something spectacular, a breakthrough to get the farmer's attention."[4]

When Harvester's corporate leadership team became aware that Murray's engineers had a decent start on the rotary combine program in the late 1960s, the executives threw money at it, hoping to take back some market share. They wrote the checks with the understanding it would be out "soon," but rotary separation development was a horrendously difficult task, and it took longer to develop than even the best and brightest expected. After more than a decade of intense development, Harvester unveiled a limited run of new Axial-Flow rotary combines for 1977.

Only three hundred units were built that year. Years of developmental work was done in secret, but the cat was let out of the bag when the dastardly competition spied them leaving the factory. Camiel Beert was a field test engineer for Harvester at that time. He had been with them since 1949 and had become the go-to

The new Axial-Flow combine was a boon to Harvester dealers like Bill Wachholz, shown here at the Bishop & Wachholz dealership near Kasson, Minnesota, on May 20, 1983. *Author's Collection*

contact when a customer's combine needed to be set up properly. Setting up a combine is a particularly tricky task, and his decades of experience sent him on trips around the globe to solve problems and ultimately convert customers of the other colors to the red camp.

"The Red Demon is climbing up on me!"
- Radio message by Deere operator

Beert recalled those first three hundred Axial-Flow combines going out the factory doors.

"We kept them all on the factory grounds until we had them all built, and when we decided to ship them. . . most all of them went by rail on a big train flatbed car and parked them [in] . . . front of our friendly competitor, which managed to drag out their cameras and their slide rules and so forth," Beert said. He added that his team recorded the spies busily checking out the new units. "We took a few pictures of that too, you know."[5]

Beert's expertise at getting the most out of a new machine would prove invaluable at that 1977 Farm Progress Show. That year, all the machinery was struggling during the demonstrations because of extremely muddy trial fields. The boys in red had a set of heavy-lugged tires designed for use in rice fields shipped in, and they mounted the hefty shoes on their new combine.

The next morning, the new combine was tearing up and down the field with aplomb, while the competition's machines—equipped with normal old tires—slipped and slid into their wake.

At the time, radios were used for the green and red crews to communicate with their machines in the field. If you happened to figure out the frequency of your competitor's communications, you could tune in and listen to what they had to say. Maybe pick up some 411.

On the day the Axial-Flow was digging mud with its big tires—while the Deere unit struggled along with everyone else—the Harvester crew was tuned into the Deere radio channel. They overheard one green operator exclaim, "The Red Demon is climbing up on me!"[6]

The wet crop plugged up Deere's combine, and the machine got stuck in the muddy ground. After those calamities, the operators were instructed to stay as far away as possible from the Red Demon.

Of course, the new combine was going faster simply because of a pair of tires; it had nothing to do with the new rotor that was the heart of the machine. That was—at least to the teams on the field and the hundreds of potential customers watching the demonstrations—not terribly relevant.

The red guys won a round that day. The Axial-Flow combine would prove to be a great new machine, especially effective at harvesting corn with very little waste. The rotary separation was a better technology than what was on the market at the time, and Harvester was able to convert some of the competitors' customers to its machines.

This drawing shows the rotor in the center of the combine that separates the grain from wheat. The technology works quite well, but it was difficult and expensive to develop. The Harvester team started on this in the 1950s and took two decades to perfect the system. *US Patent Office*

This drawing shows how the straw walker combines work. They use shaking plates to separate the grain. Deere continued to refine and use this older but effective technology on its combines until 1999, when it introduced the Single Tine Separation (STS) line of rotary combines. *Henry Dreyfuss Collection, Cooper Hewitt, Smithsonian Design Museum*

But the New Holland TR70 was a competitive machine that also featured rotary technology. Massey-Ferguson had a solid machine on the market and a loyal customer base that had been running them since the 1950s. And Deere was Deere.

The Deere combines used an older technology to separate grain, but it was a well-developed system, and the combines were smartly engineered and well-built. Some farmers said they were better in tough conditions when the crop was heavy or damp, and plenty of owners preferred the convenience of having all their equipment from the same manufacturer.

In fact, despite Harvester's new machine, Deere continued to gain market share and easily held on to the number one position in the combine market.

WARGAMES

The Twentieth Century

The Heartland

The tractor turf war dates back to the very first machines.

The first tractor in recorded history was designed by John Charter and Franz Burger for the Keystone Tractor Company. They first appeared in the press in 1887 and, reportedly, did not function terribly well, but they were put to work out in the fields of South Dakota. While these crude machines were toiling away, one John Froelich spotted them.

He promptly ran off with the idea and built his own version of the rig. By 1892, he had a workable version built, and he dubbed it the "Froelich Tractor."

In perhaps the ultimate irony, the Keystone Tractor Company would later get absorbed by the International Harvester Company, while Froelich would go on to become part of the Waterloo Boy Company, which would get purchased by none other than John Deere.

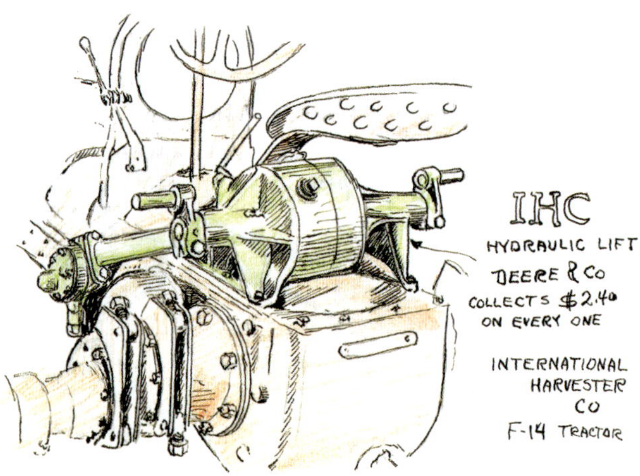

John Deere's star engineer Theo Brown made this sketch. His notes refer to the fact that Harvester's lift system on the Farmall F-14 tractor had been found to be a patent violation, and the courts ordered Harvester to pay Deere $2.40 for each lift Harvester sold. The Harvester engineers dubbed Theo "the awfulest Brown." *WPI Archives / Theo Brown Diaries*

THE IMITATION GAME

July 1972

Geneseo, Illinois

Chief engineer Don Murray got a tip about a strange-looking combine parked at a truck stop. When he went over to investigate, he was shocked to discover a machine that looked quite familiar to the one that he had dedicated much of his life to developing—Harvester's Axial-Flow combine.

What shocked Murray the most? The combine he discovered with impossibly similar technology to his was not red—it was another color, built by a competitor! The new red combine was still in development. In fact, it was four years away from being ready for production.

Peeking under hoods or driving a competitor's unattended machine are not behaviors that are tolerated in today's agricultural equipment battles—and they weren't things management was happy about forty years ago—but they were the unofficially accepted rules of engagement.

At the time, if you knew a competitor's machine was sitting unattended in a parking lot or field, well, you went and checked it out. And snapping photographs of anything you found out in open space—that was all fair play.

The game had progressed a bit from the early twentieth century battles, where sabotage and fisticuffs were too often par for the course.

One thing that is decidedly off the menu, then and now, is outright theft of intellectual property. While Harvester was developing its new rotary combine in the 1960s, one of its engineers, Bill Rowland-Hill,

When the Axial-Flow combine production units were introduced, John Deere purchased one of the early machines and gave it to development engineer Dr. Glenn Kahle. He put the Harvester rotary separator inside of a radical design of his own—this combine with a grain tank. The concept ran but was never produced. *Henry Dreyfuss Collection, Cooper Hewitt, Smithsonian Design Museum*

became obsessed with a more complex design that used two small rotors rather than Harvester's larger single rotor. Murray, the engineering team leader, felt that technology was not the way to go.

Rowland-Hill abruptly left Harvester and went to work for a rival agricultural equipment company, New Holland. At the time, New Holland had only a small slice of the combine market share. In addition to keeping an eye on the industry leader, Harvester had been closely watching Massey-Ferguson. Massey-Ferguson had taken a leadership position in combines during World War II, and it remained one of the bigger players into the 1960s. New Holland was a bit off the radar.

It was July 1972 when Murray received the fateful tip that there was an unusual combine parked at a truck stop at Geneseo, Illinois. He examined it closely—and found the only difference was that this rival machine's separating system had a twin rotor rather than a single.

When Rowland-Hill left Harvester, not only did he take the idea for the twin-rotor rotary separator design with him, he took drawings as well. New Holland developed furiously, and the machine Murray examined at a truck stop was put on the market a full year before Harvester's Axial-Flow![7]

The test crew was not above getting their hands into the machines—even to the point of running them—if the opportunity presented itself. As the engineering team learned more about New Holland's new machine—and its corporate lawyers filed suit against New Holland for the theft of the technology—one of the twin-rotor combines was located in Southern California.

Harvester engineers Bob Francis and Tom Stamp found an experimental New Holland twin-rotor machine operating in Imperial Valley. Two farmers were running it; no test people were around. "One day it was sitting out in the field, so Bob decided to

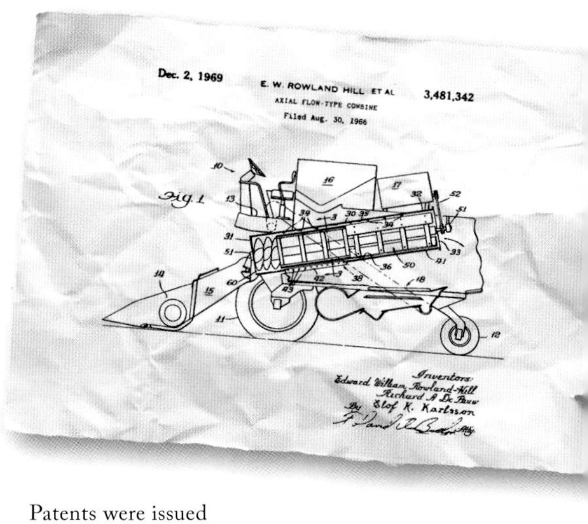

Patents were issued to Harvester engineers for the Axial-Flow combine in 1969. *US Patent Office*

investigate it," Francis said. "We found it would run so we cut a half tank or so of grain."[8]

Engineers being engineers, decided that running the competitor's prototype machine, while helpful, did not really tell them what was going on.

So, they took it apart.

The New Holland TR70 combine was introduced in 1975 and was the first mass-produced combine to use rotary separation. New Holland opted for a separation system that used two rotors. The combine was powered by a Caterpillar V-8 engine. *New Holland*

"Then I thought we should see what was inside, so we went to work and soon had a couple of concaves out on the ground," Francis said. (Concaves are the housings for the rotor that separates the grain.)

Not surprisingly, the owners of the farm eventually became aware that trespassers had not only driven their top-secret test machine—they were busy disassembling it.

"Then, here came the two farmers wanting to know who we were. Tom was tempted to just leave quickly and say nothing, but I said 'No, we will have to fess up,'" Francis said.

"They were quite upset. We left our names, tried to visit a little, and left."

Of course, the farmers reported that they had been spied on by some Harvester folks. In a matter of a few hours, word had traveled through the New Holland ranks.

Francis reported the incident to lead engineer Don Murray, who had a few minutes to digest matters before word traveled from New Holland to the Harvester executive and legal team—none of whom were amused.

"One day it was sitting out in the field, so Bob decided to investigate it," Francis said. "We found it would run so we cut a half tank or so of grain."
–Engineer Bob Francis

The New Holland leaders threatened to sue, among other things. "It all calmed down in due time," Murray wrote. "Of course, IH management was not too happy with 'that reckless bunch over there at East Moline.'"[9]

While the rebel engineers survived that day, the issue of who stole what in regard to the New Holland twin-rotor combine ended in a legal battle between the two corporations that settled out of court.

Some openly acknowledged that a good high-profile court case—ideally one with lots of lurid headlines—was an effective marketing tool.

This was not the first or the last time the two companies resolved patent issues with legal battles—court cases had been a part of the game since the Deerings and the McCormicks did battle in the 1800s. In fact, early harvesting companies considered legal fees simply part and parcel of doing business. Some openly acknowledged that a good high-profile court case—ideally one with lots of lurid headlines—was an effective marketing tool.

Patent battles have continued to be part of the game, but the tractor rivalries of the late twentieth century had a genteel air. An excellent example of this occurred when Harvester built its limited run of three hundred of the brand-new Axial-Flow combines. (All three hundred were, incidentally, hastily assembled at the factory in a space that became known by the plant workers as the "Looney Room" because building a rotary combine is horrifically complex.)

Deere, of course, wanted intimate knowledge of the new machine. While the first three hundred were sold only to hand-selected Harvester customers, Deere was able to persuade one of the customers—a farmer from Minnesota—to part with his new machine. Lore has it that the deal was a trade for a brand-new green combine as well as a big bag full of cash, but only the farmer and the Deere dealer know for sure.

What is certain is that Deere sent the new Axial-Flow to one of its lead engineers, Dr. Glenn Kahle. He and his team tore it up, analyzed all the flaws, and proposed a new Deere design built around a rotary separator that used a host of radical and wildly expensive new tech (Deere declined to build that).

When Harvester released an update for the combine—a package of new parts designed to fix some of the inevitable flaws found in a brand-new machine—one of the Deere engineers called the chief of Harvester's combine engineering team, Don Murray, and asked him to send over a copy of the operator's manual and any other factory literature that would be helpful.

Don wasn't happy about it, but he sent the update.

The John Deere man then asked him to send over a test engineer to help set up the machine.

Don declined that one.

Not long after, the factory offered an update package and Don thought it was polite to call the folks at Deere and inform them about the upgrades. He did, however, make them drive over and pick it up themselves.

Competition was competition, but the people who worked for the two companies often lived in the same communities, or at the very least, ran into each other at engineering society events. The gamesmanship was tempered with civility.[10]

The Axial-Flow combine was developed in secret in a locked building with blacked-out windows that only a few were allowed to access. *Wisconsin Historical Society / 114736*

MR. WATT'S NEIGHBORHOOD

1970s

Moline, Illinois

The red and green combine engineering teams both had significant presence in Moline, Illinois, and could keep close tabs on each other. Engineer Don Watt remembers taking carefully calculated strolls on a nearby public walking path that happened to run high on a dike, one with a convenient view of the Deere plant.

"About every two or three weeks, I'd go and walk by on the dike just to see what the hell's going on," he said. He was walking near the factory, not the engineering works. When he spotted a lot of activity—he recalled a large number of air-handling devices on one occasion—he would take notice.

"Okay. They're doing something in the plant. What the hell are they doing?"

Watt is a machine head, which is why he ended up in engineering.

"I was born and raised on a dairy farm. Always loved machinery. Hated cows and I didn't like getting up at five in the morning to get a dirty cow's tail in my face. I really didn't like that. My father was very much the opposite. He loved cows. Machinery was there as a means to the end. My brother was halfway in between.

"I decided that when I finished high school, I was leaving home, going somewhere, and I eventually went to the University of Guelph in Guelph, Ontario, and didn't know why I was there, but I knew I didn't want to be at the home farm. The first year was basically a general course and then you had to decide in what direction you were going to go. I chose ag engineering with the power machinery option."

> "If there was anybody that knew what everybody else was doing, it was Deere and Harvester."
> – Engineer Ken Ryan

This led Watt to a long career designing combines, first for

Artists like drawing wild concepts nearly as much as farmers enjoy spotting prototypes in the field. Some of the drawings, like this one by Keith Burnham, closely resemble the production machine that resulted (minus the half-cab glass doors, of course). *Keith Burnham*

Harvester's Canadian arm before he ultimately transferred to the US, where he became the head of the combine engineering team. He spent most of his life decoding the signals that would come in about the opposition's machinery.

Watt said that once you knew something was going on, there were a lot of angles to work to figure out exactly what was going on.

Companies that build heavy equipment use a lot of the same vendors to source components like pumps, hoses, and other common items. The reps from those vendors can have information. The same is true for independents who work for a variety of different makers.

"You talk to suppliers. You talk to contractors who are working there."

Ken Ryan, a longtime International Harvester employee who started with the company in the 1950s, noted the fact that many of the company employees lived and worked in the same area made secrets tough to keep.

"I had a friend that started at Deere the same time I started at IH, and we would get together on weekends. Of course, he could never get into Hinsdale, and I could never get into where he worked," he said. At the time he started, Deere was developing its all-new line of tractors, the New Generation. "I kind of had a hunch that something was happening, even though [the New Generation tractor] was one of the best-kept industrial secrets ever."[11]

When the new tractors appeared, Ryan said, "I don't think any of us were too surprised," adding that "if there was anybody that knew what everybody else was doing, it was Deere and Harvester."[12]

Watt added that while they didn't always have precise specifications on a competitor's new machine, they typically were aware of the machine's existence—and often around when it would be introduced. "They're going to launch a thing in Kansas City in September," Don said, adding that the key to good sleuthing was keeping your nose to the ground in search of the photographs that have always preluded the launch of any next big thing.[13]

In the 1970s and 1980s, these clues wouldn't show up on social media and forums, as they do today, but they did show up, with photographs and notes taken by all sorts of people circulating in the offices and boardrooms of rival companies. Technology aside, the common denominator is that customers love the game as much as (or more than) the employees.

"You can do whatever you want until they run you away. They do the same with ours. I know they do."
– Engineer Don Watt

"Farmers, they just love to tell everybody 'I saw a John Deere prototype combine.' They like to take a bunch of pictures. They'll pull into a truck stop and by God, there's two trucks with big, big combine pieces on it. They take a bunch of pictures of it. You try not to have that happen, but it happens."

Truck stops are often gold mines, and truckers are often good sources of intel. Like the contractors and vendors, one truck driver might work for multiple manufacturers—and they were frequently willing to talk about an unusual job. Don gave an example:

This sketch by HDA staff dated June 5, 1968, blends late 1960s hot-rod ethic with the New Generation's dominant impact on the industry. *Henry Dreyfuss Archive, Cooper Hewitt, Smithsonian Design Museum*

"You hauled a big combine?"

"Oh yeah. It was a big son of a bitch. It took two trucks to haul it."

If the measurements are unusual—especially if they are extraordinarily high—the engineering team knew it was likely something new.

Sometimes, they could get intel on when and where a rig would be parked overnight. "If we find them at a truck stop, we'll photograph the crap out of them and measure them with a tape measure. At least until somebody runs you away. It's out on public property. You can do whatever you want until they run you away. They do the same with ours. I know they do."

Others would try to catch the machines out working in the fields, but Watt said that was rare and difficult to do. The machines were typically very, very similar to production models, and only the most experienced people were able to notice the little differences in the details.

"It takes a pretty trained eye to see what's unique and what's different, but if somebody grabs a bunch of pictures, and of course we've got marketing guys and service guys all over the country.

"There are a number of people who are a lot better than I am at that, but they just have this mental image of what it was. They're like, 'Oh. That's different.' You're like, 'Oh yeah. You're right. It's different.'

"It takes a lot of skill and a little deception every once in a while."

JOHN DEERE

EYE OF THE TIGER

Late 1970s

Chicago, Illinois

The genteel rivalry between the tractor makers—particularly between Deere and Harvester—took on a sharper edge as the 1970s wore on. The losses of the previous two decades were grating on the old giant, and the debt load carried by the company was a serious strain.

Deere, in the meantime, had ramped up its spending on research and development, and it was in a much stronger position. Harvester needed to gain some share in order to right the ship.

Engineer Ken Ryan had been working for Harvester since the 1950s, and his career matured with a company that was an absolute powerhouse.

"They had a steel company in Wisconsin. . . . They had all their own bolts—IH bolts—they had their own bearings. All these materials IH manufactured.

"So when you were designing something or building something, you could use standard IH materials and parts for so much you wouldn't believe it. The company had their arms around everything they needed.

"Everything was done right. We were using stream gauges and brittle coding way before anybody else in the machinery world.

"We just had a really outstanding group of people. IH was strong in design and engineering, they were strong in their manufacturing skills."[14]

As they began to see the signs of trouble in the 1970s, it was hard for longtime staff like Ryan to wrap their head around it.

"How can we be losing out? We've got our act together, we're building a great product, how could this thing be slipping downhill? So it

In the early part of the twentieth century, John Deere was one of the world powers in agricultural equipment, but it was a distant second to the huge conglomerate that was Harvester. Its tractors were known for being smartly engineered, stone reliable, and stylish. *NARA*

(Left, bottom) Harvester was formed as a conglomerate in 1902 and led the industry for six decades. The Farmall tractor was its hallmark farm machine. Released in the early 1920s, it was the first tractor able to plow, turn a belt, and cultivate. *NARA*

William Hewitt Deere (left) headed the company from 1955 to 1982, and he was the catalyst that drove Deere to take the lead in the agricultural equipment industry. He is shown discussing the new Deere headquarters with designer Eero Saarinen in 1956. *Yale University Library*

was hard for us to watch it all happen."

Harvester's dominance extended to more than just numbers. It turned its capital into supply chain domination, purchasing mines, steel plants, ships, twine plants, lumber mills, and vast tracts of forest.

But as the years passed, Deere steadily gained traction and chipped away at Harvester's market dominance. Company leadership was exceptionally careful with the nickel. New products were developed thoroughly—for the most part—and the company established itself as a very solid number two to the monopolistic Harvester empire.

And Deere was always more profitable. Year over year, David came closer and closer to overtaking Goliath. When Bill Hewitt took over the reins at Deere, he declared they would no longer settle for second fiddle. His engineers, hidden away in a butcher shop, built an entirely new machine that they introduced in 1960. By 1963, Deere was number one.

By 1966, Harvester had a miniscule edge in tractors, holding 23 percent of the US market compared to Deere's 22. In combines, Deere had surpassed them, holding 35 percent compared to Harvester's 25 points.

Harvester desperately needed to stop the slide if it was going to stay in the game. A lifetime of living on thin margins and two decades of erratic choices had buried the aging giant under a mountain of long-term and short-term debt.

His engineers, hidden away in a butcher shop, built an entirely new machine.

The company had put a McCormick back in the top leadership slot in 1971, a year when Harvester's sales topped

$3 billion for the first time. The last McCormick to lead the company was Fowler in the 1940s and early 1950s. A stylish and worldly man, Fowler was a visionary who saw that Harvester had to expand into new markets to grow. Unfortunately, he was a bit of an absentee leader and his choices for expansion were, well, terrible.

The other McCormick leader of the era, Harold, had performed reasonably well until he divorced his first wife, the daughter of John D. Rockefeller, and married a young and flamboyant Polish actress . . . which prompted Harold to have glands transplanted into his body to improve his, er, performance. This was all made very public—and while the story was that the glands were extracted from a monkey, the word on the street was these glands were purchased from (and removed) from a man of modest means (the buzz was "hobo").[15]

When this story became the hot gossip of the country, as you might imagine, Harold's leadership role in the company was greatly reduced and he was shuffled off to an "advisory" role. Perhaps somewhat snake-bit from its experiences with Fowler and Harold, Harvester had shied away from McCormicks in leadership roles for two decades plus.

When Deere released its New Generation tractors for 1960, it had finally developed the technology to overtake Harvester. The new machines were a leap forward in styling, power, and features, and Deere took the number one slot in the industry in 1963. *Library of Congress*

Brooks McCormick changed that, mainly by being an upstanding and admirable human. He came up through the ranks and was a hard-working and intelligent leader. A bit of a press darling, he took over as CEO in 1971 and quickly began making the right moves.

He shuttered unprofitable divisions, changed management structure, and started funneling money toward much-needed research and development. The Axial-Flow combine was one of the results, and it proved an able weapon in the battles of the day. The technology was indeed better than the existing combines, and Harvester was able to wrest back a few of the customers it had lost to green and to other colors. While it perhaps didn't take as big a bite out of the competition's market share as Harvester hoped, it gave them a much-needed boost of sales and a hot new technology to tout.

By 1972, the roles of the two companies had switched. Deere was the number one maker of agricultural equipment and Harvester was firmly in second place. *Wisconsin Historical Society / 11319*

The company would need more new machines to make meaningful gains. The tractor line in the 1970s, in particular, was in need of improved transmissions, higher horsepower, and greater operator comfort. The machines were a good-looking lot, but the tech was long in the teeth.

The competition in this era was particularly fierce. Deere had followed the 1960 introduction of its New Generation tractors by pouring the coal on research and development. It added more power with the 20 Series in the mid-1960s and released the "Generation II" line for the 1973 season.

The red brigade had built some great machines that answered these calls in the 1960s and early 1970s. The problem was that Deere was doing that as well—and doing it with more power and better technology.

When Brooks was amping up the aging Harvester empire to gain

Brooks McCormick took over the leadership reins at Harvester in 1971. The highly regarded executive was a descendant of the founder; he pushed to modernize the aging giant and drove investment in new machinery. *Wisconsin Historical Society / 120412*

back some ground, the company had spent a solid decade falling behind the steady stream of new machines pouring out of the Deere plant.

The tractor line was outdated, and Harvester began the long process of building an all-new high-horsepower tractor in the mid-1970s. Even with massive investment, that line would take at least five years to emerge.

Harvester patched the hole a bit with some fresh paint and a few gee-whiz features—a technique it had used all too often in the late 1940s and early 1950s. It rolled out the "new" 86 Series in 1976. The machines were a good-looking lot and stone-axe reliable, but they lacked the new technology demanded by the farmers of the day.

THE PROFIT PROBLEM

1970s

Chicago, Illinois

One of the issues that Harvester battled in the 1970s had roots that extended six decades deep. While Harvester and Deere were both founded to build agricultural equipment in the 1830s, the two companies entered the tractor and powered equipment market on completely different tracks.

The International Harvester Company (IHC) was formed with a massive merger in 1902, one that gave the new megalith more than 80 percent of the agricultural equipment industry. IHC was ruled in unruly fashion by two families, the Deering and the McCormicks, who spent nearly a decade squandering their competitive advantage with petty squabbles and didn't actually come together as a single entity until the court ordered them to do so.

Even with all the infighting, backstabbing, and general mayhem, IHC had more money than anyone else in the agricultural equipment industry—plenty to dump some into the wild new frontier that was tractor development. It entered the market cautiously in 1906. By 1911, it had several engineering teams building tractors and had become one of the dominant makers in the industry.

The McCormicks had also pretty much outwitted the Deerings, and the McCormick family would provide leadership of the giant firm until the 1970s.

John Deere founded the Deere company, and his descendants would run that company until the 1980s. The merger that created IHC left Deere the odd man out. Its line of implements and dealer network was strong, just not as strong as 80 percent of the competition residing under one roof.

The Deere firm had to play its cards very carefully in this era, as the monopolistic IHC could have quite possibly run it out of business. Careful with its funds, Deere waited to enter the tractor market until 1918, when it purchased the Waterloo Boy Tractor Company.

By 1975, when this photo was taken in Nashville, Tennessee, Harvester's long history of low margins was causing it serious pain. A large debt load had built up, and customers were demanding more horsepower and newer technology. *Library of Congress, LC-DIG-ppsss-01067*

In that year, John Deere held a 12.2 percent market share of total farm implement sales with annual sales of $39.9 million[16] while IHC commanded 59.3 percent of the market with $204 million in sales.[17]

In a trend that would continue, Harvester's profit margins were slim. In 1918, Harvester's profit was only 7 percent of its sales, while Deere's was 14 percent. Harvester survived on thin margins for the bulk of its history—typically half as much as the 7 percent recorded in 1918. This lack of profitability contributed to it losing the top slot and hampered its ability to compete in the equipment wars of the 1970s.

In 1975, Harvester president Brooks McCormick made no bones about the company's error: "We became a sales-oriented organization, assuming the more volume we had, the more money we would make. We simply did not put sufficient emphasis on profitability."[18]

By the time of McCormick's tenure, Deere's decades of playing David to Harvester's Goliath had paid off, and Deere's sales exceeded those of Harvester. With aging plants and a smoldering heap of debt, Harvester had to make the right moves in the 1970s to right the ship and compete.

McCormick made great strides, racking up record sales and tripling profits—and tightening the reins on how the company tracks performance along the way.

The quest for profits would define how McCormick structured a deal to one of his key hires—a deal that would play a role in the fortunes of the entire company.

E NEW
JMBER
RNATIONAL
• 5288 • 5488
RIDE THE SPIRIT OF '76
Our special edition lawn tractor with a patriotic streak
INTERNATIONAL HARVESTER
Colfax County Implement Co.
Get a free International Scout
It's all part of the IH 4 plus 4 deal.
4plus4deal
AGRICULTURAL EQUIPMENT
NTERNATION 2+2 TRAC
2-wheel performance...plus 2-wheels more dri
Meet the all n Inter
Bringing you unparalleled fiel in the 130-150 hp range
consin teel Sold
West Coast Firm Pays $65 Million
STEEL WORKS
The Daily Calumet
Reveal Energy Plan Details
Lower Ride Prices
Cops Nearly Sure Of Dead Girl's ID
Second Front Page Eight
Vote On Inland Contract
$645 Given To Fire Fund
THE RED EDGE
2+2 ROW CROP TRACTORS
No others like them anywhere
79 H 3
SEC. ROW SEAT
BALCONY $6.50
BOSTON GARDEN
GOOD ONLY WED. EVE. MARCH 17 1976
WEATHER HERMAN CLASSIFIEDS
THE NEWS
THURSDAY NOV. 18, 1976 PAGE 47
City Edition
2,800 Workers Here Strike International Harvester

Chapter Two

THE SNOOPY SAGA

1976 - 1979

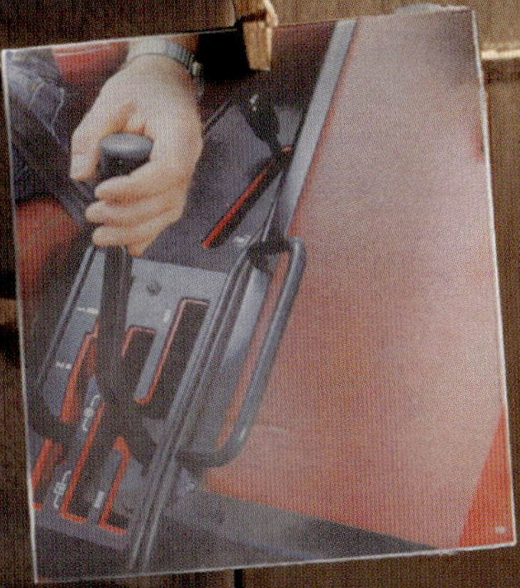

DAWN OF A NEW MACHINE

May 1976

Chicago, Illinois

As a cold dawn rose over the Harvester corporate headquarters at 401 North Michigan Avenue in Chicago on May 19, 1976, chairman and CEO Brooks McCormick was basking in the glow of news reports touting two strong quarters of sales. As the day wore, the sun would warm the air of the windy city to a balmy seventy-six degrees, and readers across the country were treated to the good news that Harvester's loan balances had been reduced and its sales were at record levels.[19]

A week later, on May 25, the Chicago weather cooled off with lows in the thirties, and Deere released quarterly numbers that took the wind out of the Harvester sails. Deere reported record sales of $900 million, up 47 percent from the previous year—and $265 million higher than Harvester.[20]

McCormick was pushing Harvester in the right direction, but so was Deere's leadership. This was bad news for everyone else, and the other colors would lose share in nearly every single market segment of the agricultural equipment industry.

This Scout II is one of less than a dozen such trucks that were built for the 1976 US Olympic Ski Team. This one is shown with an Evinrude snowmobile. *Wisconsin Historical Society / 118248*

This is an experimental version of the first 2+2 tractor built in 1976. The new model would take three years to develop for production. *Wisconsin Historical Society / 129109*

These were times of change for Harvester, and the initiatives McCormick had been putting in place since the early 1970s were beginning to take effect. Just that year, the company had shut down production of light trucks—a hard decision that helped the company focus on its more profitable medium- and heavy-duty trucks.

It also decided to sell the aging Wisconsin Steel, which was a convenient in-house supplier of steel, but one that required massive amounts of cash to keep upgraded.

Shedding or shutting down these properties would enable Harvester to fund the creation of new hardware and inject life into its aging product lines. Harvester's popular sport utility truck, the Scout, was getting a much-needed upgrade, a new plant for the Solar Division was funded to produce more turbine engines, and the newly formed Component Group was getting cash to pursue a new 6.9-liter diesel engine.

McCormick had pushed through a new corporate structure—something at least the analysts considered to be a vital change and a difficult pill for a hidebound old company like Harvester to swallow.

An initiative that was even more vital to the company's future was better long-term planning, and McCormick was the driving force behind a number of committees that were formed to study the markets and look farther ahead than Harvester had in the past.

Several objectives came out of the committees' findings. One was a long-term plan to introduce an all-new line of two-wheel-drive machines. This would take six years to accomplish—a fairly typical amount of time when it comes to developing new heavy equipment models.

The other project was a rabbit that Harvester pulled out of a hat; it used existing parts to create an odd-looking but innovative four-wheel-drive tractor. The machine emerged from a 1976 Harvester study of the four-wheel-drive market that discussed the unexpectedly rapid increase in the demand for high-horsepower four-wheel-drive tractors in the early 1970s.[21] The report opens with a painfully detailed summary of the company's failures in this market—including eye-opening figures that showed that a full 24 percent of the engines in the International 4366 had experienced "catastrophic failure."

Note that the reason for the engine failures was a common one in the heavy tractor industry at this time. The DT466 engine in the 4366 was good, but it was designed for a truck. Truck's pull heavy loads when accelerating, then are under a 50–60 percent load when cruising down the highway. Heavy tractors are under 100 percent load for much more of their lives. (When tilling heavy soil, for example, they are under full load almost the entire time.) Even a good truck engine, under this kind of strain, will fail. Harvester was not the only manufacturer to learn this the hard way—most of the manufacturers experienced the same situation.[22]

The DT466 was capable of powering the International two-wheel-drive tractors—the 966 and 1466 for example—but the heavier 4366 proved too much for the engine as configured. The engine was modified during early 4366 production, and the revised version was able to last, but early on, the engine was a disaster.

Harvester's line of heavy four-wheel-drive tractors—including the 4366—was built by Steiger out of North Dakota. Steiger built the "Cadillac" of four-wheel drives. The tractors it built for Harvester used a lightened Steiger chassis and a variety of Harvester components as well as the red paint and IH badging.

Note that Harvester specified ⅜-inch steel for the 4366 chassis. The Steiger tractors used heavier ½-inch steel. The International 4366 also used axles and differentials from the 86 Series tractors, while

The Harvester heavy four-wheel drives were contract-built by Steiger. In 1976, plans were in place to upgrade the line with a new 400-horsepower Steiger-built model. *Wisconsin Historical Society / 129090*

Steiger used much beefier axles built either by Raba—the Hungarian heavy equipment supplier—or by Steiger.

These differences meant the Steiger was four thousand pounds heavier and better able to pull heavy loads. Kelly Kravig was a marketing manager with Steiger in those days, and he recalled laying out chassis parts on the factory floor for customers to show them the difference. They also took them out in the fields and compared the machines side by side.

Deere continued to develop new equipment at an unprecedented pace. This model was built in 1977, and the work done on it would result in another high-tech new line of machines that debuted in 1982. *Henry Dreyfuss Collection, Cooper Hewitt, Smithsonian Design Museum*

The Steiger models of the day were more expensive than most of the other four-wheel drives—and had the quality componentry and build that earned them the moniker, the "Cadillac" of four-wheel-drive tractors.

Even with the lighter componentry, the Steiger-built International tractors were good for opening fields and other heavy work, but they couldn't work narrow rows and they lacked a power take-off (a spinning shaft on the back that powers implements).

And then there was the small matter that the IH-built engines, at least in the first few years of production, tended to implode.

The big Steiger-built IH tractors also tended to work better with larger implements. If a farmer wanted to upgrade to one of these big machines, he was handed the bill for both an expensive, new high-horsepower tractor and a pile of spendy implements large enough to take advantage of the big machine's power.

The bright spot in this rather glum document was that it identified an opportunity in the four-wheel-drive tractor market. At the time, high-horsepower two-wheel-drive tractors were popular with large producers who needed a machine to work with the narrow crop rows of the day. Narrower crop rows put more plants in the field, which meant a higher yield per acre.

There was an opportunity here if someone could provide a light, four-wheel-drive machine that fit the bill. The gem in the document is a photo of a nicely finished prototype for a new Harvester-built tractor that could do just that.

In his typical style, Bud Youle summed up the competitive situation: "Nobody in the late seventies—nobody—had a four-wheel-drive tractor that was acceptable in the marketplace. The one we had for the 86 Series looked like two drunks and an alcoholic put it together."[23]

He also knew that John Deere didn't have a platform in its existing line that could easily be modified to offer the adjustable spacing and maneuverability required for the new narrow rows. Its front-wheel-assist row-crop tractors were solidly built, but the wheels were fixed. The big four-wheel drives were also solid units—albeit with a less-than-ideal engine—and the row width would not come down.

Harvester's idea could offer adjustability. The concept that the engineering team cooked up was to mate two rear-end drive units from the 86 Series two-wheel-drive tractors together, hinge the new machine in the middle, power it with the venerable 466 engine, and build a maneuverable four-wheel-drive tractor. They dubbed the new machines the 88 Series, or the 2+2.

Using two rear ends was key, as the wheels could slide on the axles. This meant the machine could be adjusted to work narrower rows quite easily—unlike the machines of that dastardly old competition.

The model needed an estimated three years of additional development, particularly to get the tooling to build them in place as well as to finish the design of the fairly complex hydraulic system

The machine was also odd-looking. This was not a positive attribute. The original Farmall concept was nearly dustbinned entirely in the 1920s because it was "odd-looking." When demonstrated in 1920 to the

> "Nobody in the late seventies—nobody—had a four-wheel-drive tractor that was acceptable in the marketplace. The one we had for the 86 Series looked like two drunks and an alcoholic put it together."
>
> – Bud Youle

executive team, the machine would move forward and backward and looked more than a bit like a mechanical crab. The executive's reaction to the decidedly oddball contraption was close to revulsion. Longtime executive Alex Legge—who was an advocate for the concept—famously pleaded with his engineering team leader, Bert Benjamin, to at least make the Farmall look *something* like a tractor.[24]

Harvester was also working on a new line in the late 1970s. This is a 1976 drawing for its new high-horsepower two-wheel drives, which would be introduced in 1981. *Gregg Montgomery*

Farmers and executives alike are reluctant to open their pocketbooks for bizarre new concepts.

Engineers John J. Bardos and William L. Schubert wrote that the reaction from the farmers they recruited to test the 2+2s was fairly consistent. "Most people would comment somewhat disparagingly about the tractor silhouette, particularly pointing out the long nose of the tractor. They were sure the long hood would be a problem. However, as soon as they climbed aboard and started to operate the tractor, they were completely convinced that the concept was correct."[25]

The fact that it was distinctive looking also meant testing and transport had to be done with the utmost care. If anyone spotted an early unit in the field, they would be able to spot it was a new concept from a mile away.

The whole development project was hush-hush, top secret.

"Very few people in the Harvester company knew we were building this tractor. I mean, it was we'd take it out on field tests, we had our own truck, we had our own people," Youle said. "It was such a well-kept secret that we would stand at crossroads, and they'd count the cars and farmer's trucks that go by, before we would go to Mr. Smith or Mr. Jones's place to put it in the field to run it."[26]

The funny-looking machine continued to test well, giving the red faithful hope they could take a slice out of the green tractor people's pie.

THERE'S A NEW SHERIFF IN TOWN

August 1977

Chicago, Illinois

As Chicago's downtown was pelted with rain and steamy eighty-degree heat in August 1977, Harvester was again in the news. The company sold off the aging Wisconsin Steel for $65 million (in a shaky deal that would haunt it later).[27] The sale made headlines on August 3. A day later, big changes at Harvester generated more coverage.

On August 4, 1977, Brooks McCormick revealed that he had recruited a CEO to replace him. He chose Archie McCardell, the highly touted second-in-command at Xerox. McCardell cut his teeth as an analyst and director of finance for Ford Motor Company. At Xerox, he oversaw record revenues and profits in the mid-1970s (never mind the fact that the company was stumbling mightily when McCardell exited).

In 1977, Brooks McCormick (right) recruited Archie McCardell from Xerox to take the helm of Harvester, paying him the highest salary of any executive in the US at the time. *Wisconsin Historical Society / 120324*

McCardell's hiring made headlines, both because Harvester's leadership had historically been groomed from within and for the fact that McCardell's $450,000 salary made him one of the highest-paid executives in the US.[28] The deal was sweetened with a $1.6 million signing bonus plus a $1.7 million loan at 6 percent . . . the loan being forgivable if McCardell raised Harvester's profitability to be equal to or higher than the average of its main competition: Deere, Ford, General Motors, Paccar (who made trucks), and Massey-Ferguson.

McCardell's $450,000 salary made him one of the highest-paid executives in the US.

Historically, Harvester had the industry's lowest profit margins. One can imagine the hiring committee approving the personal loan believing the task so close to impossible that it was a very safe bet. Obviously, if he did it, the company would be wildly profitable and, if not, he would have to pay them back—plus interest.

McCardell was highly incentivized to deliver both big sales and big cuts in expenditures. The engineering team found this meant they could get money when they had ideas. McCardell appears to have understood how critical it was for Harvester to get some new technology on its dealership floors. Among the VP ranks, a running joke emerged that McCardell had a "money tree" in his backyard, and if you had a new idea for a product that might help Harvester move forward, the tree would bear fruit and provide ample funding.[29]

Under McCardell's tenure, new technology was funded at an unprecedented rate, leading executives to joke that he had access to a "money tree." One radical experiment was the Advanced Harvesting Systems: a group of engineers and scientists recruited to develop a harvester—the Probine—that could process crops into various solid and liquid proteins rather than simply separating the grain. The concept did not pan out. *Gregg Montgomery Collection*

THE STAR-SPANGLED SALESMAN

1978

The Heartland

Salesman Bud Youle would spend a good part of his career talking to farmers, either helping them solve problems or showing them the latest machines. *Amy Santi Collection*

When Harvester's ace young hotshot John "Bud" William Youle first laid eyes on the 2+2 tractor, the company was closing out another year of record-setting sales. The leaders of that pack were the truck group, whose new line had been a hit. Farm equipment had racked up solid sales, with strong results from their new combine and heavy four-wheel-drive tractors, results that were unfortunately offset by a sluggish farm market that slowed sales of their aging two-wheel-drive tractor line.

The machine and Youle were both well-oiled new weapons in Harvester's battle to thrive—and, for that matter, survive.

A native of Hannibal, Missouri, Youle had paid his own way through college and, in three and a half years, walked out with a degree in economics.[30]

He served four years commanding a platoon in the Marines. "The proudest day of my life was when I held up my right hand and took the oath legion to protect and defend the constitution of the United States—so help me God—and was sworn into the marine corps," Youle said. "I loved every minute of it."[31]

The military made its mark on Youle, and he carried the whip-like bearing and the buzz cut of a leatherneck for the rest of his life.

"I believe in doing things the way George Patton did. Head that army up and move them out and don't stop until you run out of fuel, and when you do, load your guns and start shooting," he said. "That was my philosophy of business. I loved it."[32]

"Being a warrior in the marine corps has helped me throughout my life," he said.[33]

When the military wanted to send him to work in Okinawa, Youle was reluctant to go and leave his wife and children behind. While home in St. Louis, he decided to check his options.

He was able to arrange a meeting with Fred Fernald, a district manager for International Harvester who was based in Kansas City. Youle was hired on the spot. He went back, got his discharge, and two weeks later, reported to work in KC.

"The Harvester company and God Bless America have been my bywords. I very much devoted my life and my family to the things that happened in the Harvester company. And I was very, very fortunate to meet some great people in the Harvester company," Bud said.

"I worked with some that used to get so mad at me, they would throw pencils at me." He also said he had thick skin, so the pencil throwers never bothered him much.[34]

A gifted storyteller and commanding presence, Youle quickly distinguished himself for his ability to connect with farmers in the field. He solved their problems, filled their ears with stories, and brought them into the company fold by sheer force of personality.

On the first day on the job, his new boss, Jim Garner, called him into his office.

"Bud, you've got to go run a forage harvester. Nobody else in the office knows how to run one," Jim said.

Youle headed out to Princeton, Missouri, to run that forage harvester for a dealer with a colorful nickname. ("His nickname was Sip and Sigh and he loved bourbon whiskey.")

> "I believe in doing things the way George Patton did. Head that army up and move them out and don't stop until you run out of fuel, and when you do, load your guns and start shooting."
>
> – Bud Youle

5,000,000th
INTERNATIONAL
HARVESTER
TRACTOR
FEB.1.1974
5 MILLIONTH TRACTOR
INTERNATI
1066
FARMALL

Youle quickly found that relationships were his best tool when out in the field helping dealers and customers run red machines.

"I learned real quick the two simple basic principles of business: One, people do business with people and, two, a handshake was worth a million dollars," he said. "I never lost those two traits. If I shook hands with you, a deal was a deal."[35]

He cut his teeth as a zone manager, traveling around to visit the dealers and farmers in his region. He drove town to town in his pickup truck, spreading the word about new red machines. "I knew every farmer in that part of the world," he said.

Some of them were easier to deal with than others.

One of those he remembers well was a farmer near Herrington, Kansas, who kept burning out clutches on his tractor.

"Bud, there's a farmer out there, big farmer, tough, mean ornery," Youle recalled being warned. They were going out to replace the clutch in his Farmall 806 tractor—for the second time.[36]

Bud and the dealer hopped in the truck and headed over to Ewald Kickhaefer's farm to see why the clutches kept failing in his relatively new tractor.

"Ewald was a corporal in the German army in World War I, so you could see what I was confronted with already right off the bat. But anyway, when parked—I parked there at the house—we walked around a couple barns. Now all of you now remember this, there was no cabs in those days, and old Ewald was driving that 806 up the field with that left foot on that clutch pedal."

Youle noted that Kickhaefer rested his foot up on the clutch pedal as he drove rather than leave it sitting on the floor. Farmers commonly do this—the issue is, if they depress the pedal, they will damage the clutch. Kickhaefer was a big man, and his beefy leg "resting" on that pedal was pushing it in plenty, causing the clutch to wear out quickly.

Needless to say, Kickhaefer was not happy with his brand-new tractor's drivetrain imploding on a regular basis.

In 1974, Harvester produced its five millionth tractor. The head of Harvester marketing—and Youle's boss—Stanley Lancaster (center) was photographed with the machine on February 1, 1974. *Wisconsin Historical Society / 155815*

"You here to fix my clutch for the second time?!" Youle recalled him saying.

"Ewald, we are going to fix the clutch today," Youle replied. "How are we going to do it Ewald? I'm going to cut your goddamn leg off right there!"

Youle's in-your-face communication style played well with the old farmer, and the two military men hit it off.

"Ewald always called me 'Blockhead' and that's when I knew I had a friend," Youle said. "To this day, that family still runs red equipment."

Stanley Lancaster recruited Youle to come to Chicago. "OK Stan, I'm your man—back to Chicago I go." So, Youle and his family moved from Liberty, Missouri, to an apartment in the Windy City.

Harvester's technology was a bit long in the tooth in the mid-1970s, but the old giant's supply chain was incredibly integrated—it made its own bolts, owned ships and airplanes, and the list goes on. This image is dated February 16, 1977. *Herbert Starzinger Collection*

He moved on to become a disciple of the construction division. “They had a habit of calling me back into Chicago. I got into Chicago, and they said, ‘We got a job for you,’ but nobody really wanted this job and I understand why they didn’t want it. I said, ‘Sign me up coach.’ I signed up for anything, no matter what it was.”[37]

> “At the time, in the regions I had responsibility for the sales department, service department, parts department, distribution department, and the bitching department.”
>
> – Bud Youle

While he worked construction, he reported to two different people: one was Patrick Kaine, who was heading the ag division at the time, and his second boss was the head of the construction division. Youle spent time working in Kansas, Pennsylvania, and Minnesota, among other places.

“At the time, in the regions I had responsibility for the sales department, service department, parts department, distribution department, and the bitching department,” Youle said.[38]

Compared to his fellow salespeople, Youle was young and brash. Whether he moved around so much because of his terrific results or because of his in-your-face office demeanor is open to debate. “I had a district manager, a regional manager that you always noticed. He sat right beside of me. He never left my side,” Youle said. “I couldn’t get him to agree on anything, and he and I locked horns many a time and it later got me in trouble.”

His manager called him into his office one day.

“You’re done, you’re toast. You’re out of here,” Youle recalled being told.

“What do you mean, you fired me?”

He said, “No, you’re going to Chicago!”[39]

The odds on whether Youle was being fired or moved were pretty much even. He said of his longtime boss, executive Stanley Lancaster, “He fired me twice, cut my pants off twice—he and I got along real good.”[40]

This was in 1976, and this time Youle was sent to Chicago to stay. The "new" 86 tractors had just been introduced. "The 86 Series tractors were very good, reliable tractors, but what was happening in the marketplace was that John Deere was on the move."

The competition wasn't the only problem. Youle was assigned to work with the Farmall Plant, and he found himself in the middle of a mess of epic proportions.

The combines had a host of problems that needed fixing. The tractors needed an all-new transmission that shifted better and would handle more horsepower. The company was trying to get its German branch to build a new compact tractor. The list went on.

Times were tough, money was tight, and all the programs needed attention. "The Harvester company was starting to become under a lot of pressure," Youle said. "Believe me, if you don't have any money, but if you've got good people, you can make it happen."[41]

Youle liked to exaggerate, but he was also a natural-born leader not to be trifled with. He knew how to fire people up. He also loved to tell a good story and wasn't afraid to add liberal amounts of exaggeration with some outright bullshit sprinkled on top to drive his point home.

> "What do you mean, you fired me?"
> – Bud Youle

One of his colleagues said that after Youle made sales presentations at regional meetings, his group would have to get together, figure out how much of what he said they could actually do, and get to work on development or damage control.

That didn't change the fact that customers and colleagues loved to be around Youle. He was a spark plug, and his high-octane sales engine would need to fire up to help power Harvester through the challenging times ahead.

In 1978, the Harvester marketing team cooked up the Red Power Showdown—events were held to inject some life into flagging sales of its 86 Series tractors, among other things. *Ron R. Schmitt Collection*

RED DAWN

January 1979
Scottsdale, Arizona

"It is the ugliest thing God ever made."
– Bud Youle

Late in 1978, Bud Youle was taken into the bowels of Harvester's high-tech development center at Hinsdale, Illinois. The engineering team had a new concept to show him—their top secret four-wheel-drive concept—the 2+2 tractor.

Youle wasn't impressed.

"It is the ugliest thing God ever made," he recalled saying. "All they would say to me is, 'Bud, we are going to have to sell this.'"[42]

The tractor's odd appearance earned it a variety of uncomplimentary nicknames, most of them keying off the machine's long nose. "Snoopy" and "anteater" were two of the most popular terms.

The Harvester crew had spent the year reading less-than-encouraging headlines about how their company profit margins didn't match the competition. Executive Keith R. Potter said the company's share of the large tractor market had dropped 10 percent because of a new line of tractors introduced by Deere.[43]

To rub salt in that wound, in November a US district court ruled that Harvester had copied Deere's corn head patent. A hearing loomed to decide how much the embattled old giant would have to pay its competitor.

By December, McCardell was touting Harvester's additional $100 million in cost cuts as well as increased investment in research and development. He acknowledged this wasn't enough to right the company ship. "We missed some market share targets in fiscal 1978, particularly in trucks and tractors," he said.[44]

With the company fortune on the line, the pressure on engineering to build something new and sales to get that new tech into the fields was immense. Ugly or not, Youle knew he had to bring home the bacon with this new beast.

"I got to come up with a way to make that tractor sell."[45]

What the Harvester team had learned, and what Youle keyed off, was that the best way to sell the odd-looking 2+2 was to put a person behind the wheel.

What he cooked up was a grand introduction to be held in Arizona in January, when most of the farmers were out of the fields and glad for an excuse to go somewhere warm.

"We are going to bring every dealer, every dealer salesman in there, and they were going to see this tractor. Now they were going to not only see the tractor, they were going to drive the tractor, because the biggest problem we had to overcome was getting people in the seat to convince them that an anteater wouldn't fall apart. It wouldn't turn and hit 'em. It wasn't going in the ground. The brakes would stop, and the hood wouldn't fall off every time."[46]

To do this, Harvester needed thirty to forty of the new 2+2 tractors built and shipped to Arizona in giant boxes (so the competition wouldn't get wind of the program). Youle claimed that Deere placed a camera crew on the grounds of a college near the Farmall Plant to take photographs of what came in and out. "I didn't want them to see what was shipping out of that plant."[47]

Bud figured the program would cost at least $1 million.[48]

As Youle tells it, he first pitched the program to Stanley Lancaster, who was vice president of marketing. Youle recalls him sitting and taking notes.

"The next day Stan says, 'I want you in the office at seven o'clock.'

"I walked in at seven o'clock, we

Harvester corporate headquarters was housed at 401 North Michigan Avenue in Chicago. Bud Youle would go to the twenty-sixth floor to make his pitch to roll out the 2+2 tractor in Phoenix. *Library of Congress*

went upstairs. We didn't stop at the eighteenth floor, we went to the twenty-sixth floor."

"'Stan, we're going way on up.'

"'You are going all the way on up,'" Youle remembered Lancaster replying.

> "The biggest problem we had to overcome was getting people in the seat to convince them that an eater wouldn't fall apart."
>
> - Bud Youle

The twenty-sixth floor was the executive level, and Youle said there were several heavy hitters from Harvester in the room. He said at various times that Brooks McCormick, Archie McCardell, and others were present as he made his pitch.

"I said, gentlemen, you give me the million dollars and I'll give you six thousand orders two weeks after the program is over."[49]

His pitch worked. The money tree was shaken down, and one million bucks was approved to fund the 2+2 launch. Harvester got behind the funny-looking machine the competition had dubbed as "Snoopy"—with guns blazing.

"I knew that nobody in the Deere company could conceive what we were doing. That was the only guy that I was really setting my sights on," Youle said.[50]

"I knew that if I could make this tractor sell in the south, we could keep . . . the cash coming."

In the hopes of keeping the balance sheet in the black, Harvester used Bud's one million bucks to make the Snoopy show a first-class affair. It rented Dick Van Dyke Studios. The machine rolled out on stage in the morning, accompanied by dancers, music, and every bell and whistle that money could buy.[51]

The Harvester team took the dealers to lunch at the Pinnacle Peak Patio, a western-themed steakhouse in Scottsdale, and loaded them up with mesquite-grilled thirty-two-ounce steaks, western kitsch, and ice-cold Budweiser.

"Then, that afternoon we took them to the field," Youle recalled, adding that the test fields were east of Scottsdale. "Every man, woman, and child that came to that meeting had to drive these tractors on this course."

"Now, the thing was when they looked at it, when they saw this tractor, oh god, we knew that was the thing we had to overcome. You have to take your worst aspect and make it your best feature. That's what we did."[52]

They also had to do a little damage control. The test units were shipped with the sliding hoods in place—but the rubber straps to hold the hood in place were missing. That meant that under hard braking, the hood could slide right off these early production machines. Bud had to prep his test crew to make sure no one made that mistake during the demonstrations.

"Now I told every guy in the field, 'You look 'em right in the eye and you tell 'em this: This is the ugliest thing God ever built,' and you tell 'em that when they're going seventeen miles an hour and slam on the brakes and that hood's going to fly right off the front end of that tractor."[53]

Each region was brought in together—all the dealers from the territory centered in Kansas, for example, would arrive at the same time. Youle

The 2+2 tractor launch party included a meal at the Pinnacle Restaurant (above). This is a similar western-themed joint in Scottsdale: the Handlebar J. *Carol Highsmith, Library of Congress, LC-DIG-highsm-53380*

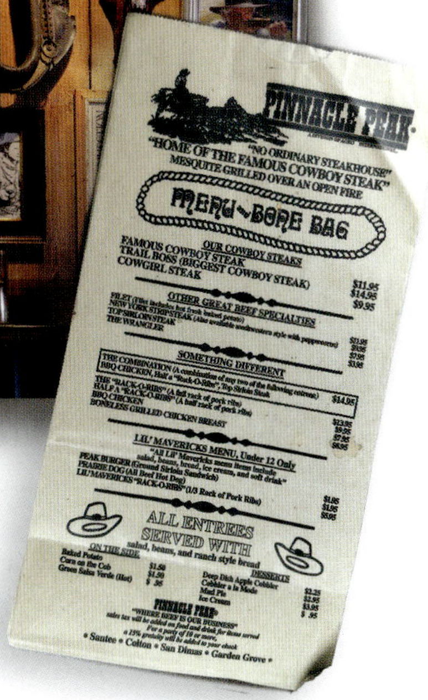

The Pinnacle Restaurant specializes in steaks (menu on right), and dealers who flew into Scottsdale for the 2+2 launch in January 1979 were treated to a meal there. If a guest showed up at the restaurant wearing a tie it was promptly cut off. *Wikimedia Commons*

"'You look 'em right in the eye and you tell 'em this: This is the ugliest thing God ever built.'"

– Bud Youle

counted on them talking to each other, reporting to other regions what they saw. "Dealers have a communication system better than the CIA," he said.

The first regional dealers Harvester flew in were from the Memphis area: dealerships in Alabama, Mississippi, Louisiana, Arkansas, and Tennessee. Youle was worried about them because of the John Deere 4840, which was a helluva good tractor that was popular in the region and was rated to produce 185 horsepower.

After the dealers were done driving the machines, Harvester had them watch a demonstration in which the new 150-horsepower 2+2—model 3588—faced off against a John Deere 4840. The engineers at Hinsdale had a 4840 they were using to test, which was shipped to Scottsdale for the trial. In addition, Youle said they knew the 185-horsepower tractors were being set up with 200 horsepower by the Deere dealers, so they did the same.

The 3588 2+2 was down a full 50 horsepower in the contest. The machine would compete with traction not power.

Bud drove the machine himself, to make damn sure it performed its best.

"This wasn't a game; it was serious business," Youle said. "The destiny of the Harvester company laid right there with that tractor."

The hopped-up Deere and the long-nosed Harvester machines were matched side-by-side in the field, pulling with all they had on a short field.

"We set up a skit. Now, my cohort in this skit was . . . Chuck Lare." Youle said. "He was the villain."[54]

To further emphasize Lare's villainy, the team dressed him in a John Deere green jacket and shirt, topped off with a green and yellow John Deere hat.

"We hooked to offset disc curls and when we said go, the tractor started up the field pulling these offset disc curls."[55]

Red and green machines faced off in the field. The winner was the first one across the finish line. "I did that ten times a day for every region," Youle said. "I probably won ninety percent of the events."

The 2+2's four-wheel drive put power to the ground and allowed it to rule the day. Dealers also saw the new 2+2 working narrow rows and turning quickly.

The demonstration was a head turner, and Harvester finagled orders for 850 tractors out of the boys from Memphis.

After the dealer introduction, Harvester's two 2+2 models—the 3388 and 3588—were demonstrated around the country. The machine looked odd, but driving it often convinced skeptical farmers to plonk down their cash. *Ron R. Schmitt Collection*

Harvester's Farm Equipment Research and Engineering Center (FEREC) at Hinsdale, Illinois, was state of the art when built in the 1960s, and it was where most of Harvester's technical innovation was done. FEREC became known as Burr Ridge postmerger, and it continued to be a center of new design until it was closed in 2024. *Adam Summerlott Collection*

Planes came in and out all week, with dealers flying home with full bellies and empty pockets. Harvester brought in not only the North American dealers, but also dealers from overseas.

“I don’t care if you like it, you hate it, cuss it. I don’t care what you do. This tractor did things for the American farmer that nobody yet to this day [had] ever done.”[56]

“I don’t care if you like it, you hate it, cuss it. I don’t care what you do. This tractor did things for the American farmer that nobody yet to this day [had] ever done.”

–Bud Youle

“I remember I walked into Stan’s office after the show was over,” Youle said. “‘Sir, I placed orders at the Farmall plant for sixty-five hundred tractors.’ We wound up building, in the neighborhood, thirteen thousand of them.”[57]

The 2+2 was introduced to dealers at a gala rollout held in the Dick Van Dyke Studio. *Gerry Salzman Collection*

The 2+2 tractor was one of the bright spots for Harvester. It was a creative use of the technology it had at hand, and it became a strong-selling tractor in 1979 and the early 1980s. *Wisconsin Historical Society / 129161*

SEASONS IN THE SUN

December 1979

Chicago, Illinois

The new 2+2—and an absolutely booming farm economy—would boost Harvester's agricultural equipment sales in 1979. For the first time in company history it exceeded $3 billion, helping push global sales to a record-breaking global $8.4 billion and record profits of $370 million.[58]

Harvester's profit ratio for 1979 was one of the highest for the company in recorded history.[59]

Those numbers would prove critical to McCardell, who had been given a $1.796 million loan to purchase sixty thousand shares of Harvester stock as part of his signing package. The loan could be forgiven if, in a seven-year period, he was able to raise Harvester's lagging profitability to the average ratio of six competitors: Deere, Caterpillar, Massey-Ferguson, Paccar, Ford, and General Motors.

This was a tall order for the historically low-profit Harvester. At the close of 1977—McCardell's first year—Harvester recorded a 6.76 percent profitability and its six competitors average 10.5 percent. Total forgiveness of the loan prior to the seven-year mark was possible if McCardell achieved spectacular results.[60]

For farmers like Ramona Minisee—shown driving a tractor and baler near Cassopolis, Michigan—1979 was a ray of sunshine. Crop demand and prices were high, and the future looked bright. *NARA / USDA*

The profit Harvester recorded in 1979 was indeed spectacular. It is also worthy of scrutiny. First, some unusual tax credits from the United Kingdom "added" $49 million in "earnings" to the numbers for Harvester in 1979.

Second, labor relations at Harvester were publicly strained in the fall of 1979. The contract between the company and its labor force was set to expire on October 1. As early as February 1979, rumblings were heard in the press of the possibility that negotiations could lead to a strike, as they had during the last round of contract negotiation in 1973. As the year wore on, the drumbeats grew louder.

Anticipating a strike, and confronted with tremendous demand, dealers stocked up more heavily than normal. How much this affected sales is open to debate—some Harvester executives suggested this bolstered sales by 5 percent, while outside security analysts suggested this raised sales by 30 percent.[61]

In the 1970s, farm protests in the form of tractor drives to Washington, DC, were a growing trend that generated headlines. The 1977 protest took place in the winter and ended on a genteel note when a snowstorm buried DC. Several more took place in 1979, including one in February (pictured here) and another in December. *NARA / USDA*

Funny math included, Harvester's 1979 profitability was calculated at 9.55 percent. The average of the competitors: 9.15 percent.

"We put a cog right smack in Deere's plans with the 2+2," Youle said. While that may have been true, Deere was still way ahead of Harvester in terms of having good new machines in development, not to mention in the fields. Harvester also had a balance sheet that was bleeding red with debt, while Deere's debt load was much more manageable.

To make matters worse, the winds that blew favorably for Youle's company into the summer and early autumn of 1979 were about to change. Led by McCardell, the new leadership team at Harvester was determined to win concessions from the UAW.

Heavy-handed negotiations led the UAW to strike. At noon on November 1, 1979, thirty-five thousand workers from twenty-one of Harvester's plants walked off the job. This put 36 percent of Harvester's work force out of commission.

As 1979 waned into the holiday season, the joy of Harvester's dream year was dampened by an increasingly harsh reality. Its troubles would be magnified by the fact that more difficult times lie ahead.

"We put a cog right smack in Deere's plans with the 2+2."
–Bud Youle

Chapter Three

HERE COMES THE RAIN AGAIN

1980

UNITED STATES MARINE CORPS
Tempo
Daycare centers catching on — slowly
1980: tough on
Deere profits dip 27 percent
IH red ink
Tractor Performance
With This Brand
INTERNATIONAL AGRICULTURAL EQUIPMENT
International
De zwaarste
Hoe zwaar het werk ook is: International heeft de traktor. Deze bijvoorbeeld: de IH 1486. Een zwaargewicht van 122,5 kW/167 pk. Technische perfektie op maat. Komfortabel en veilig. Betrouwbaar en bedrijfszeker als elke International. Ga eens praten met uw IH-dealer. Hij zorgt voor een goed advies en voor vertrouwde service.
International
Voortrekkers in technische perfectie
NIEUW
Boeke-Heesters bv
How to Ride The Bull
COUNTRY STYLE Magazine
The STATLER BROTHERS—
All-American Small Town Heroes
HANK WILLIAMS, JR.: Exclusive Interview with a Survivor
FREE Country Dance Lessons Inside
GEORGE JONES: Back on Top and Sober
A Tribute to CHET ATKINS
Plus—
•JOHNNY LEE
•JUNIOR SAMPLES
•CHARLY McCLAIN
Inside—STATLERS & BARBARA MANDRELL
Color Photos, Suitable for Framing

ANOTHER BRICK IN THE WALL

January 4, 1980
Washington, DC

The mercury started the day below freezing in Washington, DC, and would barely creep above that on a gloomy January 4, 1980.[62] By the time the sun had set, the city was blanketed with two inches of snow and a grain embargo was declared on the Soviet Union. Much like the white flakes that slowed traffic to a crawl in the US's capital, the embargo would take the wind out of the sails of the farm economy.

The storm extended all the way east to the snow-covered cornfields of Ely, Iowa, where Maurice Garant was repairing his tractor and contemplating how to deal with grain bins full of the fall's harvest.[63]

The solitude of the farm takes its toll, one that would become more prevalent as the difficult economy of the 1980s sent droves of small families into bankruptcy. Reverand Aaron Ronken, known as the flying pastor, is pictured counseling a farmer in Montana. *Los Angeles Times photo by Joe Kennedy*

As the 1980s dawned, times turned harder for people like this farmer who raised gamecocks near Leakey, Texas. *Marc St. Gil / NARA / EPA*

That year, land prices fell. Interest rates—which had been rising throughout the late 1970s—skyrocketed. Wheat prices dipped and recovered.

All the complex array of factors that govern our economy were pointing south. Issues had been building for a long time, and the causes were myriad. The bottom line is this: the embargo that was put into place on that cold January day—a frosty dawn to a hard year—marked the beginning of the farm crisis.

Steiger employee Roger Silkey vividly recalled the day as a line in the sand between good and bad times. "I remember sitting in a meeting and [the grain embargo] was announced and the high prime interest rates were in effect," he said. "We cut our schedules in half that day."[64]

This polarizing event would kick off a year in which Harvester got to bask in the glow of its record-setting 1979 sales for about twelve minutes. The strike at Harvester plants that began November 1, 1979, was still in full force when 1980 dawned.

THE COTTON CLUB

1979-1980

Yuma, Arizona

Harvester was battling to create a new machine for the cotton farmers of the world. Cotton is one of the trickiest crops to harvest; removing the fibers is so difficult that it was done by hand long after machines were developed to harvest simpler crops, such as corn, beans, and grain.

The red machines had been leaders in the cotton harvester market since Harvester introduced the first effective cotton harvester in the 1940s. That machine took two decades to develop and, for a time, was the dominant machine in the industry.

Deere had entered the market in 1950. By the time the two makers were at it hammer and tong in the 1970s, Deere had whittled away its red competitor's lead, and the two were battling for the lead spot in this niche of the agricultural industry.

In 1914, the United States grew two-thirds of the world's cotton, which is an extremely labor-intensive crop to harvest. During World War II, labor shortages had cotton farmers clamoring for a mechanical solution. Harvester had been developing this technology since the 1930s, and it introduced the first successful mechanical picker in 1942 (but it could not produce it in large quantities until steel restrictions were lifted in 1948). *NARA / USDA*

In 1979, both manufacturers were putting the finishing touches on new models that would be introduced in 1980. The competition was fierce, and any whisper about what was going on behind a competitor's door was welcome information.

John Deere territory manager Dale Johnson was knee-deep in the battle in 1979. He was based in Yuma, Arizona, when he received a tip about a competitor's machine out in the field—in Arizona's Box Canyon.

Arizona is fertile territory for machine development and testing, as the temperate weather makes for ideal ground conditions even when the middle of the country is buried in snow. Through much of the twentieth century, new equipment was moved as quietly as possible and tested as discreetly as possible in that part of the country. To this day, the major manufacturers have proving grounds in the region.

Since the earliest manufacturers began testing new equipment there, skullduggery was afoot, with engineers and others sneaking around during the dead of night to peek under tarps, crawl under machines, take metallurgy samples, and even partially disassemble equipment.

When Johnson picked up his phone that day, he received a tip from a colleague at the Deere factory in Des Moines, Iowa, that builds cotton pickers. The Deere plant was about to release a new four-row cotton picker, the 9940, and they had a tip to pass on.

"Rumor has it that there is an International four-row experimental cotton picker working around out there some place in southwest Arizona or southeast California," the voice on the phone said. "Do you suppose you could track it down?"[65]

Johnson had a hunch he could.

"I knew who to call, my John Deere dealer in Parker, Arizona, Bill West. Bill was a grand old gentleman and knew everybody in the countryside."

> "We threw a couple of shotguns in there and I had my camera."
> – Dale Johnson

Parker was right on the California border and, as Johnson predicted, West had a hunch how to get a look at the competition's new machine.

"You know, I think I got an idea," West said. "If they're going to run something, there's a real loyal IH guy and he's in the proverbial Box Canyon. There's only one way in and one way out. We've gone hunting there."

The canyon was just across the California border, a few hours north of Yuma. Johnson didn't waste any time. "I drove up there to Parker." He was joined by Bill Weber, a salesman for West, and the pair loaded into an ordinary pickup truck. "We threw a couple of shotguns in there and I had my camera."

When they arrived in Box Canyon, which was near the Colorado River, they spotted a big red machine picking cotton in the distance. They decided to get closer to get a good look at it, which was when the cavalry arrived.

"The next thing we saw was two pickups come roaring down our way. Dust was flying!" Johnson said. The two trucks skidded to a stop, blocking Weber and Johnson's vehicle in the front and rear.

"What are you guys doing here?" the Harvester man asked.

"Well, we're scouting out hunting for bird hunting this fall," Bill said.

"What are you doing with that camera?"

"Oh, we like to take pictures too."

"Bullshit! You guys are from John Deere, aren't you? Get the hell out of here!"

The two men hightailed it out of there and went back to the dealership to consult with West.

The mission wasn't over, as West had another idea.

He owned a four-wheel-drive dune buggy with a big engine that he thought was just right for the job.

"You know what," he said to his salesman Bill Weber, "Why don't you take Dale out in the dune buggy, and you can just go through the desert and get on top of that canyon, and I'll bet you can find a way to slither down the canyon wall and get over there."

So, Weber and Johnson beat their way off-road to the top of the canyon.

"We found kind of a drainage, a crevice in the canyon wall, and so we slithered down that canyon wall and got to the bottom to the edge of the cotton field," Johnson said.

"We used some of our old army skills and we low crawled between the cotton rows and toward the cotton picker and it happened to be just right at noon time and so they had stopped the cotton

> "Why don't you take Dale out in the dune buggy, and you can just go through the desert and get on top of that canyon, and I'll bet you can find a way to slither down the canyon wall and get over there."
>
> – Dale Johnson

picker under a clump of cottonwood trees along the irrigation canal."

The Harvester test crew was sitting under trees having lunch, and Johnson was able to sneak over and photograph the experimental cotton picker. He climbed on top of the machine and inside it as well.

By the late 1970s, Harvester and Deere each held roughly half the cotton picker market. The key to their success was larger machines that covered more rows, and both companies had introduced four-row pickers (like this one). *Case IH*

"We low crawled back out of the cotton rows and climbed back up out of the canyon, and it was a lot easier going down than it was going up, I remember that."

On the rim, they discovered a deer or elk skeleton.

"We propped up the head and I took some pictures of that too, and so we called it the skull espionage," Johnson said.

The photos were sent to Deere headquarters, and word got back that they were passed around to a large number of employees there, eventually finding their way to the CEO's office.

Deere introduced the 9940 four-row cotton picker, and International put out its Model 1400 four-row picker. Both were released for the 1980 season—the two companies were aggressively attacking the cotton picker market.

"That was pretty tough competition back in those days, and so we were both trying to gain a foothold. We had probably sixty percent market share on cotton pickers, on the two-row cotton picker, and so they had the rest. There were only two cotton picker manufacturers, Deere and IH. They had probably about forty percent and we were wanting to keep growing in that market share," Johnson said. "It was pretty intense."

HARD TIMES

Spring 1980

Detroit, Michigan

The intense competition between Deere and Harvester was taking place on the precipice of disaster in spring 1980. As the two companies fought for share, the situation on the farm continued to erode as prices fell and interest rates climbed stratospherically.

With a recession looming, the prime rate was adjusted down a tiny bit in April 1980. This was a blip—the prime rate would continue to rise as the worst economy in modern history would sweep across the US and the globe.

This was bad news for anyone and devastating for heavy equipment manufacturers in particular. The down market reduced sales, and the high interest rates further depressed sales by making loans on new equipment inordinately expensive. The farmers and those selling them equipment were also kicked in the teeth with the fact that crop prices were coming off their 1979 surge and heading into the cellar.

In addition to the challenges of the times and the competition, Harvester had been in a knock-down, drag-out

The massive UAW strike by Harvester workers that began on November 1 idled thirty-five tousand workers. In April 1980, the two sides finally settled. These workers were leaving after working the night shift at Harvester's Melrose Park plant on April 21, 1980. Only a skeleton crew showed up for this first shift, but more would return as word of the settlement spread. *Sun-Times / Jim Klepitsch*

battle with its union labor force—those employed at the UAW—and more than a third of its workers had been on strike since November 1979. The strike by the Detroit-based UAW would finally be settled on April 20, 1980.

Archie McCardell, Harvester's leader at the time, said the strike may prove to be "one of our best investments" on the day that it settled. Granted, he admitted the length of the strike was too much.[66]

The gains were marginal at best. The key concession Harvester hoped to get from the UAW had been the addition of mandatory overtime, which would help Harvester ramp up production when demand was high. Both Deere and Caterpillar had mandatory overtime in place.

Harvester did not gain mandatory overtime.

The cost was certainly high. When the dust settled, Harvester lost $222 million in the first quarter. Analysts at the time were projecting the total cost for the year could be $500 million and said Harvester's best hope for 1980 would be to break even.

In order to have any chance to do that, Harvester needed to get its workers back building machinery—ideally something new.

Accomplishing that would prove nearly as problematic as the strike.

In May 1980, Mother Nature contributed to the chaos in the country when Mount St. Helens erupted and sent volcanic ash around the globe. This farmer was haying in volcanic dust near Moxee, Washington, in early June 1980. *Author's Collection*

THE BAD SEED

1980

Hinsdale, Illinois

Harvester had introduced a new planter model early in the 1970s and had held a large percentage of the market with the machine.

In the early part of the decade, Harvester was content to rest on its laurels with the machines it had. Deere being Deere, it invested in the market, put some good machines out, and Harvester found itself in a familiar situation—losing market share.

As the 1970s waned, Harvester had stepped up its R&D in this area and had a new machine ready for production. With its factory workers striking, of course, production of pretty much anything was down to a crawl, so the new planter had to wait. In April 1980, the workforce was back, and Harvester was in desperate need of some new product.

The new machines would be designed in the high-tech Farm Equipment Research and Engineering Center (FEREC) in Hinsdale, Illinois. Opened in 1959 on the site of Harvester's long-time testing grounds, the facility was state of the art at that time and referred to simply as "Hinsdale." Sprawling over eight acres and housing 1,360 staff members, the complex included an indoor "field" where full-sized implements could be tested, a photo studio big enough to house a tractor, and temperature-controlled rooms able to expose machines to temperatures ranging from 130 degrees to 50 degrees below zero.[67]

One of the people working at the facility, Steve Warner, recalls the chaos generated by the planter project. Warner started his career at Harvester in 1980. He began as an engineering intern in early April. "When I started here in April, the strike was still going. My first couple of weeks here, I went down and assembled planters on the assembly floor because the union

> "They tried to ram-rod the new concept through."
> – Steve Warner

guys were on strike. . . . We were panicking to put planters together to get them out to test them."[68]

The new machine was the International 800 Series Early Riser Cyclo Air Planter. The new technology offered by the planter was great; the machine, however, was not adequately tested.

Harvester's 800 Series Early Riser Cyclo Air Planter was a leap forward when it was introduced and was especially well suited to the growing trend of minimizing soil tilling. It was another slick piece of technology that emerged due to R&D investment from McCardell's "money tree"—it was also rushed to production and early units had serious issues. *Case IH*

When the union workers came back to Harvester, some of them were immediately assigned to get the new planter built and ready for market. According to Warner, the job they did building the machine was impressive.

"The plant workers set up to build all those planters on the assembly line. And did really quite a great job, given that they didn't have a whole lot of time to react and were still scrambling to put all the striking workers back to work."

The planter, however, would require a multimillion dollar fix to make it right.

"[Harvester] rode the laurels of their successful planter until it was too late to react to what Deere had put in place. And then they tried to ramrod the new concept through, which they did, but they ended up paying through the nose for it and not only with fix packages, but [it] also gave that planter a bad eye, a bad reputation."

Warner's job responsibilities over the next few years would grow exponentially, and while he would get the opportunity to design new machines, he also spent much of his time fixing the 800 Series planter.

He also designed a lot of new implements that were not introduced, at least not during the Harvester days. New product that lacked the necessary funding to launch would be a trend for the company as money became tighter and tighter in the 1980s.

FIGHT THE POWER

June 1980

Southeast Chicago

By June 1980, the sluggish economy of the year was widely acknowledged as something that was going to settle in and stay for a while.[69] This wasn't news to anyone who was paying attention—employment was down while interest rates and prices were up. Even Deere found the need to lay off workers, sending seven hundred home from its plant in Dubuque and another seventy-five from Davenport.

The slowing economy was not helping Harvester, which found itself in the news that month due to a lawsuit from one of its old properties—Wisconsin Steel—rising from the dead like a zombie from the grave.

Harvester had sold it in a sketchy deal at best, and the aging beast had gone bankrupt—at least partially due to the UAW strike—and saddled Harvester with a loss of tens of millions of dollars due to the sale going away.

The hits just kept coming at Harvester in this era. The troubles extended to the Snoopy tractor line.

The original vision for the 2+2 was for a total of three models. Two of those had been launched in early 1979—the 130-horsepower 3388 and the 150-horsepower 3588. The market demanded more horsepower—a trend that had started in the 1950s and lives on yet today.

To fill that gap, the original 2+2 plan included a higher horsepower version of the machine, the 170-horsepower 3788. The tractor was the cover model for its 1980 Buyer's Guide—a powerful addition to the 2+2 line.

Normally, a 1980 tractor model is introduced well in advance of the model year. This is more of a guideline for manufacturers than a hard-and-fast rule. But one can easily surmise that Harvester had wanted the new 3788 on showroom floors by spring 1980. The strike undoubtedly fouled up that plan.

Well, in June 1980, the new 3788 was not anywhere to be seen . . . and news came that it would be delayed.

On June 9, 1980, a rather grimly worded memo was sent by Harvester

employee H. R. Simmons to the lead engineers on the project as well as the Farmall Plant managers . . . and the world headquarters.[70] The upshot of the memo was that the new 3788 tractor being tested in the fields were chewing up transmissions with all the vigor of a new puppy attacking an open box of Purina Bonz.

The new machine was scheduled for release, and the memo stated that the tractor's power train would likely fail at multiple points in the first two or three years of operation.

Bud Youle responded to Simmons in a June 26 memo: "Marketing cannot have a transmission where failures in the 1,500–2,000-hour range," adding that it was assumed this new model would be just as reliable as past models.

A meeting was scheduled for July 2, 1980, at the Farmall Plant, where the engineering, marketing, and management team executives would gather to hash out the situation.

The group suggested a wide variety of technical changes needed to solve this problem, including fixing a part of the front-wheel-drive portion that they referred to as the "sleeping pig" sub-assembly.

The problems were more or less solved, and the 3788 rolled out behind schedule and over budget.

That would prove to be the least of Harvester's worries.

Another piece of the new technology pie was the higher-horsepower 2+2: the International 3788. In test units, the drivetrain would not hold up to the increased power. The long-term fix for this problem was a better final drive and transmission, which was planned for 1984. *Ryan Roosinck Collection*

DEAD FROM THE EARS UP

Summer 1980

Chicago, Illinois

As the summer of 1980 progressed, Harvester was facing a legion of hard hits. In July, a court appeal said it had indeed infringed upon a patent for a corn head held by John Deere (the court case was a crazy one—the Harvester employee had actually gone into Deere's innovation center and had been videotaped measuring the machine).

Demand for heavy trucks was down. White Motor Company went bankrupt early in the year. Harvester's truck division leader, Patrick Kaine, predicted more would follow. "The truck makers who will survive the shake-out could well be counted on one hand."[71]

As the 1980s wore on, the farmers of the day fought to survive the worst times since the Great Depression. Farmland values would drop as much as 60 percent in the coming five years, and average mortgage rates would peak in the US in fall 1980 at 18.6 percent. This farmer is from Baca County, Colorado. *Rob Amberg*

Harvester's construction division was also facing hard times and a down market, and it closed two plants in late September, sending 1,700 employees home from the Melrose Park and Libertyville facilities in Illinois.[72]

On October 31, 1980, Brooks McCormick retired from Harvester at age sixty-three. "I read an article," he said, "about bosses who hang on for too long. . . . You know, there are a lot of guys who are dead from the ears up at sixty-three. I think it's in the best interests of the business that the new teams be given a clear track. It's time to clear the old guys out."[73]

His timing hardly matched his words. Harvester was in serious trouble, and McCormick had been the poster boy for the company's much-publicized

efforts for renewal in the 1970s. He singled out the strike as "the greatest disappointment" of his career. "We had made this wonderful recovery in 1979; the strike set us back quite a ways."

> "Some people believe Archie's being given undue credit. Many of the things that happened were Brooks McCormick's doing."
> – *Chicago Tribune*

The article speculated that if McCormick had still been the head of Harvester, the strike might not have occurred. "It's hard for me to read fifty thousand union minds," McCormick countered. "I can't read the other imponderables; I just don't know."

One day after McCormick retired—and one year after McCardell's obstinance with the union led to a massive and costly strike—Harvester decided it should forgive McCardell's loan of $1.7 million due to his 1979 performance (see chapter 2, "There's a New Sheriff in Town" and "Seasons in the Sun" for more details).

Crop prices would drop drastically in the early 1980s. More critically, the cratering land values and high production left most farmers with too much debt. For example, in Minnesota alone, more than 49 percent of the eleven thousand farmers in the state were delinquent with bank payments. These local farmers are pictured delivering their produce to a market in North Carolina in July 1981. *NARA / USDA*

The loan had been given as part of McCardell's hiring package, and it was used to purchase Harvester stock. The loan forgiveness was widely questioned, with a Harvester executive telling the *Chicago Tribune*, "Some people believe Archie's being given undue credit. Many of the things that happened were Brooks McCormick's doing." A chairman from the bank who funded the loan stated, "From a public relations aspect, it was probably not a good thing to do."[74]

The loan would not be reinstated if Harvester's performance declined. If McCardell was fired, he would need to repay the loan, but the forgiveness remained in play if McCardell were to leave the company voluntarily.

Brooks was resolute in his defense of the action, "The truth of the matter is that Archie McCardell did earn it—he earned it demonstrably."[75]

Brooks was the great-grandnephew of the founder, Cyrus McCormick. He also said that he had no regrets being the last McCormick to lead Harvester, adding that one of his two sons had shown interest in joining the company and he had vetoed that decision. The son had moved on to work in the art business in New York. "There's no room in this world for emotionalism or sentimentality," Brooks said.

Though Brooks retired from the executive forces of Harvester, he remained in an advisory role at the company.

His departure came at a time when all the heavy equipment manufacturers were suffering. In December 1980, Deere reported its first reduction in profit in a decade. Sales were above the previous year, but profits were off by 27 percent. Caterpillar had a tough year as well, reporting layoffs and sales below expectations.[76]

When it came to corporate pain, Harvester took the cake in 1980. In December, it announced $397.3 million in losses for fiscal year 1980. The gains of 1979 were effectively nullified in a catastrophic year that married a five-and-a-half-month strike with rising interest rates, falling crop prices, and a wicked economic recession.

Despite the difficulties of 1980, Brooks remained optimistic, "We continue to make progress in cost cutting and capital outlays, and next year will be our biggest ever for new product introductions."

Under McCardell, Harvester had increased research and development spending significantly, and the new machines built by the "money tree" would be seeing light the following year.

"We will absolutely beat [Harvester's performance] in 1979," Brooks said. "We just have to make more money."[77]

The Scout line was put up for sale in May 1980. When a deal with a Texas buyer fell through, the decision was made to end production of the Scout, and this photo shows a truck and dark signs on the line during the last day of Scout production, October 21, 1980. *Mike Bolton Collection*

McCormick was spot on, and the company's performance was not quite as dire as the numbers indicated. Agricultural equipment sales for 1980 clocked in at $2.5 billion. While that was 18 percent off an incredible 1979, it was a respectable showing in a down market.

To survive, the company needed to show off the work of their engineers . . . and sell as it had never sold before.

Guys like Bud Youle would be under the gun in 1981.

Chapter Four

RUNNING UP THAT HILL

1981

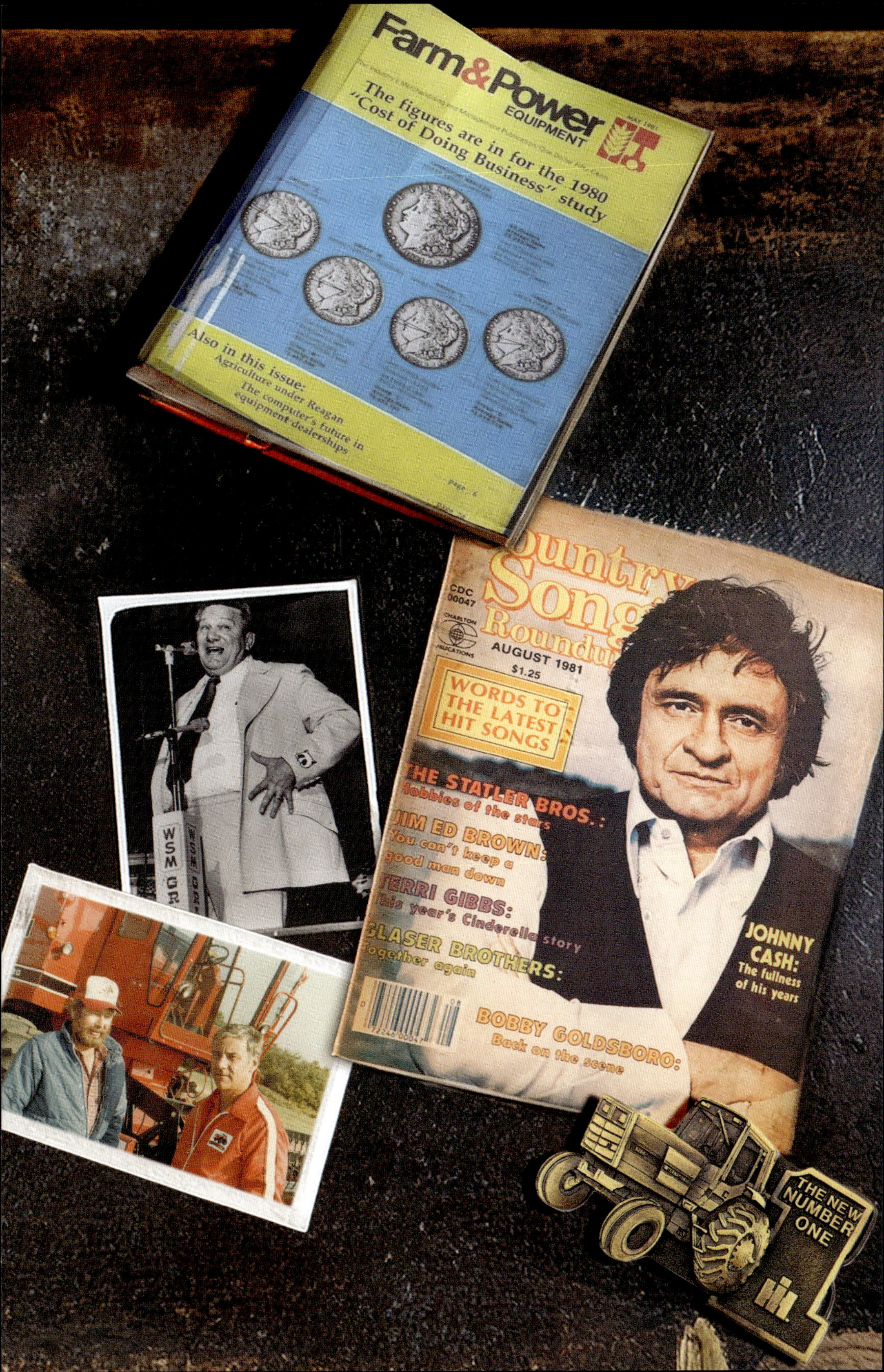
Farm&Power
EQUIPMENT
The figures are in for the 1980
"Cost of Doing Business" study
Also in this issue:
Agriculture under Reagan
The computer's future in
equipment dealerships
CDC
00047
AUGUST 1981
$1.25
WORDS TO
THE LATEST
HIT SONGS
THE STATLER BROS.:
Hobbies of the stars
JIM ED BROWN:
You can't keep a
good man down
TERRI GIBBS:
This year's Cinderella story
GLASER BROTHERS:
Together again
BOBBY GOLDSBORO:
Back on the scene
JOHNNY
CASH:
The fullness
of his years
WSM
THE NEW
NUMBER
ONE

DON'T STOP BELIEVIN'

January 1981

Chicago, Illinois

As 1981 dawned, CEO Archie McCardell continued to beat the drum for Harvester. Despite a tough outing in 1980, he touted the improvements the company made in cutting costs.

"We have been trying to attack this cost problem," he said. "We still have a long ways to go." His tone was upbeat about the product line coming—his money tree was about to bear fruit. The heavy investment in R&D gave Harvester a record number of new models for 1981.[78]

As New Year's Day hangovers wore off, Harvester's hopes began to wane. On January 15, the company announced it was cutting the pay of the company's top twenty-six corporate officers by 20 percent and freezing the pay of thirty thousand other managers and white-collar workers.[79]

A week later, on January 23, it announced that executive Warren J. Hayford's loan for $973,438—a deal similar to the one given Archie McCardell—was forgiven.[80] With heavy cuts in the news, high pay for executives did not sit well with stockholders, analysts, and the rank and file.

McCardell's money tree—the seemingly bottomless well of R&D money enjoyed by the engineering teams around the company—also finally died off. An unnamed company spokesman added that the 1980 reductions would include "slowing the rate of investment in many areas."[81]

By February, Harvester was reporting first quarter losses of $96.4 million. Its short-term debt increased by $807 million to a total of $2.48 billion. On February 3, Moody's Investors Services announced that it was downgrading Harvester's rating, a move that would make it harder and more expensive for the company to carry and obtain financing.[82] To worsen matters, the failure of the Wisconsin Steel plant sale had led to 3,500 workers being laid off—a large portion of whom placed the blame squarely on Harvester.

Seeing one's company in the headlines each day was rough on employees, and Bud Youle recalled a lot of change on that front. "A lot of people were resigning and leaving because every day somebody would say 'cashflow, cashflow, cashflow.' I couldn't spell cashflow, but I probably learned what it meant."

He added that the loss of experienced personnel made the job more complicated. "New people would come in and when new people would

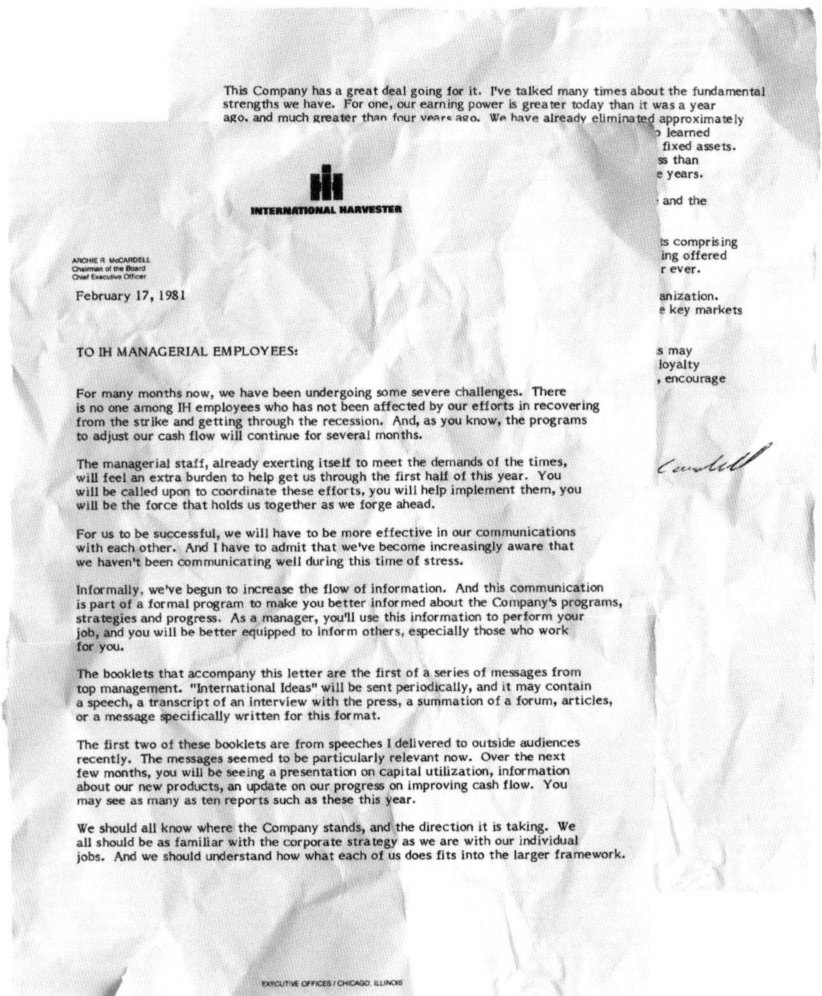

This Company has a great deal going for it. I've talked many times about the fundamental strengths we have. For one, our earning power is greater today than it was a year ago, and much greater than four years ago. We have already eliminated approximately

INTERNATIONAL HARVESTER

ARCHIE R. McCARDELL
Chairman of the Board
Chief Executive Officer

February 17, 1981

TO IH MANAGERIAL EMPLOYEES:

For many months now, we have been undergoing some severe challenges. There is no one among IH employees who has not been affected by our efforts in recovering from the strike and getting through the recession. And, as you know, the programs to adjust our cash flow will continue for several months.

The managerial staff, already exerting itself to meet the demands of the times, will feel an extra burden to help get us through the first half of this year. You will be called upon to coordinate these efforts, you will help implement them, you will be the force that holds us together as we forge ahead.

For us to be successful, we will have to be more effective in our communications with each other. And I have to admit that we've become increasingly aware that we haven't been communicating well during this time of stress.

Informally, we've begun to increase the flow of information. And this communication is part of a formal program to make you better informed about the Company's programs, strategies and progress. As a manager, you'll use this information to perform your job, and you will be better equipped to inform others, especially those who work for you.

The booklets that accompany this letter are the first of a series of messages from top management. "International Ideas" will be sent periodically, and it may contain a speech, a transcript of an interview with the press, a summation of a forum, articles, or a message specifically written for this format.

The first two of these booklets are from speeches I delivered to outside audiences recently. The messages seemed to be particularly relevant now. Over the next few months, you will be seeing a presentation on capital utilization, information about our new products, an update on our progress on improving cash flow. You may see as many as ten reports such as these this year.

We should all know where the Company stands, and the direction it is taking. We all should be as familiar with the corporate strategy as we are with our individual jobs. And we should understand how what each of us does fits into the larger framework.

EXECUTIVE OFFICES / CHICAGO, ILLINOIS

This February 1981 letter is from Archie McCardell to the executive team, urging better communication and promising a bright future. *Adam Summerlott Collection*

> "Sir, you have twenty-eight hundred and thirty-four tractors parked in my yard in San Antonio, Texas. You got one week to get 'em out."
> – Governor of Texas

come in, I'd have to sit and explain things. I mean, because they didn't know a tractor from a B-52 bomber. I'm serious."[83]

Another headline for Harvester in 1981 centered on a deal that had been made to supply pumps to the Soviet Union to be used for construction of the Trans-Siberian Pipeline. The previous administration had approved the deal, but Ronald Reagan had just taken office, and his office wasn't sure it should move forward.

Harvester had bankers, lawyers, and the competition at its throats—why not add the president of the United States to the list? Just get in line.

A foreign government and the state of Texas would soon join in. The news came in a phone call.

Youle recalls getting that fateful call while he was at Hinsdale, the Harvester research center in Illinois.

Bud picked up the line, and a voice barked in his ear.

"Yes sir, I'm the governor of the state of Texas.'"

Youle was stunned.[84]

"Holy lord, mercy for mine. What did I do now?" he asked.

"Sir, you have twenty-eight hundred and thirty-four tractors parked in my yard in San Antonio, Texas," the voice said. "You got one week to get 'em out."

"Now, how did I get twenty-eight hundred and thirty-four tractors sitting at San Antonio, Texas? I did not know."

Whether the governor actually called Bud—or whether it was the governor's office—is open to question. The number of tractors sitting in that lot in Texas is also open to debate.

But what is known is Harvester's Farmall Plant had indeed built tractors for Mexico. People who were there, as they best recollect it, think that just over 2,500 machines were built, and they were built in three separate production runs. What most likely happened

By 1981, farmers were going out of business on an all-too-regular schedule, and equipment manufacturers were offering bargain basement deals. This is John Vogel, who stayed on his foreclosed land until escorted off—peacefully—by law enforcement officers on November 8, 1986. *Bill Gillette, State Historical Society of Iowa*

The 1970s saw a huge increase in land values. Farmers faced with ever-increasing demands to produce more with less often used their newfound equity to finance some of the fantastic new equipment being developed. When land values and crop prices cratered in the 1980s while interest rates skyrocketed, the squeeze was terminal for many family farmers. This farm was owned by a farmer that received a low-interest federal loan in 1973. *NARA / USDA*

when Youle got that call was that six hundred to eight hundred of them—the final of three runs—ended up stuck in Texas.

"Sometimes the marine corps and the navy, they didn't fire the guy that made the mistake. You knew they fired the boat commander, the colonel, that's the guy that got fired."
– Bud Youle

Tractors were sitting in that lot, so the story goes, because whoever was paying for the machines wasn't able to come up with the cash. They were built specifically for a market in Mexico, with special decals, one-off tires, and minimal options. To the best of everyone's recollection, they were International 986, 1086, and a few 1486 models.[85]

This meant a cash-strapped organization fighting for its life now had a large number of tractors to get out of Texas, and Youle was the guy who had signed the production order.

"Sometimes the marine corps and the navy, they didn't fire the guy that made the mistake. You knew they fired the boat commander, the colonel, that's the guy that got fired," Youle said. "Everybody was looking for me because it took my signature to get all these special tires and all that stuff done on it."

Youle's ass was on the line. He got on the horn and started making things happen.

"I called my friend Ed McKnight and I said, 'Mr. McKnight, I want you in San Antonio, Texas, tomorrow morning at eight o'clock for breakfast. You and I got an assignment.'"[86]

The pair double-timed it down to San Antonio. What they found was a massive lot filled with tractors.

The 86 Series tractors destined for Mexico that ended up in San Antonio were sold off at fantastic prices. The tractors were basic models with few frills and unique tires, and they became known as "taco tractors" among collectors. *Author's Collection*

For farmers with older equipment and a reasonable debt load, the attractive pricing of machinery made in the early 1980s proved a fine time to upgrade. *Rob Amberg*

McKnight stayed in San Antonio for three weeks, getting trucks in and tractors out.

After the tractors were out, they were put up for sale at rock-bottom prices.

Farmers at the time were receptive. Those that had older equipment and were unwilling to drop $50,000 to $100,000 on the newest machinery were happy to get a solid, brand-new 1086 for slightly more than $20,000.

Youle spared no quarter moving those machines. "For three months, I was the meanest human being God ever put on the face of the earth," Youle said.

"I'm in South Georgia, I'm up in Georgia, I'm up in North Carolina. I'm at the Mississippi Delta. I'm at Tom Scotts in Louisiana. I'm going as fast as I can go, and Delta Airlines was getting tired of hauling me."

Double-time paid off.

"We finally got the tractors all sold out, no lawsuits," Youle said. "Everybody seemed to be happy."

THE MONEY TREE BEARS FRUIT

Fall 1981

Chicago, Illinois

As 1981 rolled into the fall, Harvester was frantically scrapping for cash. Two of its most profitable divisions, the Cub Cadet line of lawn and garden tractors and the high-technology Solar Turbine Company, were put on the block and sold. Cadet became an independent firm and netted Harvester $276 million.[87] Solar, which Harvester had purchased in 1960, was sold to Caterpillar for $505 million in cash.[88]

The company had major problems with its cash flow, but McCardell was resolute that Harvester continue to invest heavily in new tech. In February, he stated that the 1981 research and development budget would be higher than 1980's all-time record of $384 million.[89]

While Harvester kept its R&D investment for 1981 to itself,[90] Wayne Broehl put together a comparison of Deere and Harvester's R&D expenses from 1977 to 1981. According to his research, the only year that Harvester outspent Deere on a per employee basis was 1981.[91]

This investment would pay off in fall 1981, when Harvester introduced its new products—with the highlight being the all-new 50 and 30 Series tractors—to its dealers at a gala event held in Kansas City in September.

The culmination of work that began in 1976, the two new model lines featured all-new styling, a host of new technology that included a radical new cooling system, mechanical front wheel-drive, and—most significantly to the marketplace—an all-new synchronized transmission.[92]

The transmission was a compromise. The market demanded a full powershift transmission—meaning one in which the operator could change speed on the fly without stopping. An old-school manual transmission required depressing the clutch and moving the gear lever. Tractors in the 1940s and 1950s largely had this type of transmission.

For a sports car enthusiast, changing gears with a manual transmission is a gas. For a farmer, it's a drag that cuts productivity and can be hard on equipment. When loads change, the operator has to press the clutch and change gears. Often when a tractor is pulling hard—say

In mid-February 1981, the news broke that Harvester sold the Cub Cadet line of lawn tractors to a newly formed subsidiary of MTD Products Inc. of Cleveland, Ohio. The highly successful line began for Harvester in 1960. Under the new arrangement, the lawn machines would continue to be sold at IH dealerships. *Wisconsin Historical Society / 26992*

Engineer Dr. Glenn Kahle was recruited to Harvester from John Deere. He would lead the engineering department through the last days of Harvester and beyond. *Glenn Kahle Collection*

plowing—the tractor will stop dead when the clutch is engaged. After the gear change, the machine has to get back up to speed.

Time is lost, and lost time can be critical, especially if your harvesting or planting window is narrow . . . or you don't have enough people to work on your farm. Labor shortages were, and continue to be, a problem on the farm. So, the holy grail of tractor technology in the 1960s to 1980s was a transmission in which you simply moved a lever to increase or decrease speed.

Harvester had released a partial solution, the Torque Amplifier (TA), on its 1954 models. The TA gave farmers the option of changing one speed on the fly. It was great at the time, but you only had one speed change when moving. Ford had released a technology that did this in 1959, the Select-O-Speed transmission, and it was not reliable. Deere released its powershift transmission on its 1964 models, which was reliable and fairly well-received (though some complained about the fact that the transmission slightly reduced the rear wheel horsepower).

Nonetheless, Harvester was way behind with its transmission by the 1970s. McCardell's money tree was shaken down heavily in an effort to correct this, and Harvester's engineering team had a wide variety of high-tech new transmissions in development.

Most of these slick new gearboxes were projected to cost more than the company's money tree could produce. Harvester was so desperate that it approached Deere about sourcing the transmission from them. "We went out to Waterloo and talked to their engineers, and they came through with a concept [transmission]," Harvester engineer Ed Harnach said. "The cost was just prohibitive."[93]

In the end, Harvester's new-for-1981 machines had a transmission that featured three different ranges and could be shifted on the fly in each range. A tremendous improvement, to be sure, but not the full powershift the market demanded.

While the finishing touches were put on the new 50 Series tractors, the engineering department saw a change in leadership. Dr. Glenn Kahle was a lead engineer at John Deere for many years, overseeing a group that developed all-new technologies. While he was there, he was approached by Harvester about working for its engineering team.

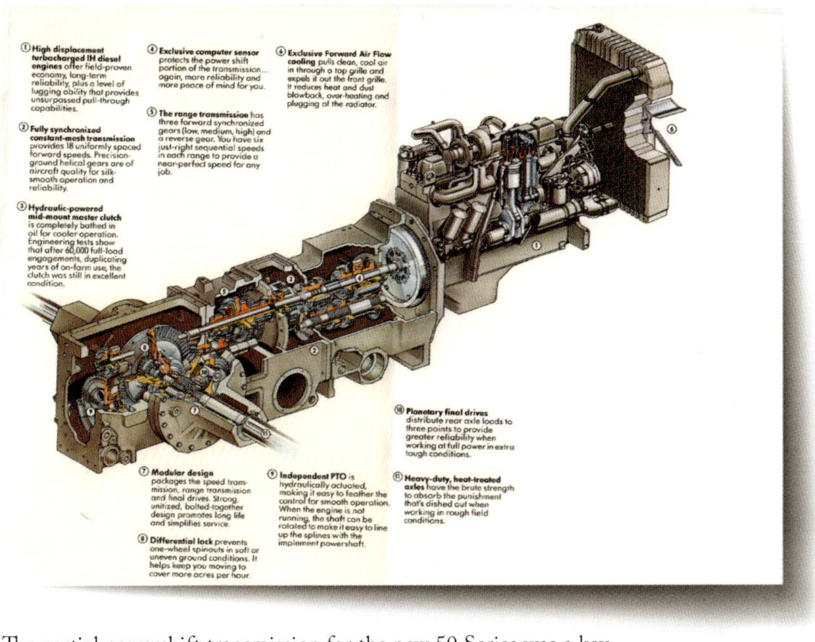

The partial powershift transmission for the new 50 Series was a key feature desired by farmers. It was expensive to develop and much-needed—Deere offered a full powershift in 1964. *Case IH*

"I was scared. I mean, you had the sense you'd get fired on the spot if anybody knew about it."
– Glenn Kahle

While engineers would occasionally leave one group and work for another, it didn't happen often. The green and red engineering camps just didn't cross-pollinate. Kahle remembers being nervous when he was approached.

"I was scared," he said. "I mean, you had the sense you'd get fired on the spot if anybody knew about it."[94]

They asked Kahle to come to Chicago for an interview.

"I was literally reluctant to even go. I remember sitting in the headquarters downstairs with a magazine up over me so nobody would recognize me.

"In Chicago nobody would anyway, but I didn't know that."

The Harvester offer enticed Kahle to join, and he became chief engineer. When he came in, the research and development money was flowing, and his group had their heads down building new technology. Kahle found the change refreshing.

"It was really an exciting, opportune time with people working as hard as we could to get things done. And of all the organizations I've been in, and that's been a half a dozen, I saw less backbiting and backstabbing and all the things that go on in the organization; there was practically none of that. To this day, it was probably the best experience of my life." Kahle said.

This is a 50 Series tractor undergoing anechoic testing at Hinsdale. *Adam Summerlott Collection*

Like Bud Youle, the Snoopy salesman extraordinaire, Kahle seemed to have relished the challenge presented by Harvester's hard times.

"There just wasn't anything on the table except

get the job done and get it down now, because the fear was they were going to go bankrupt."[95]

Harvester's new-for-1981 50 and 30 Series tractors were the result of a program that began in the mid-1970s. The new models were introduced at a gala event in Kansas City—this photo is from that event. *Roger Lewno Collection*

Kahle's first task when he started was to oversee the completion of the 50 and 30 Series machines. He said the compromises required on that model were significant because Harvester didn't have enough money. "With the IH program, we didn't have enough money to put a new cab on it, for example, and it needed one."

The cab wasn't the only aging component on the new 50 Series. The engine was a good, solid Harvester model . . . but not a new one. Kahle believes if Harvester's finances were in better order, the engineering team would have built a much better tractor—one with a powershift transmission, a new cab, possibly a new engine . . . and a myriad of details they had developed but lacked the funds to produce.

Kahle said the models would have been very similar to the Case IH Magnum, a tractor introduced using Harvester's technology in 1987. "So yes, IH would have done it if they had had the money, and it would have been a very similar tractor [to the Magnum]."

The Harvester engineering team also had designed and created early prototypes of an all-new midsize series of tractors, dubbed the TR4 models, but did not have the money to put those into production. Midsize tractors were a growing market and one that was hotly contested in the 1970s. The Harvester line of 84 Series midsize tractors was aging, and a new model would have been welcome.

Even with the company finances flagging, Harvester invested heavily in the introduction of the 50 and 30 Series. They were introduced

The team at the launch. Unfortunately, Harvester lacked the funds to fully develop some of the technology it had designed for the new 50 Series. *Roger Lewno Collection*

to dealers in Kansas City in September 1981, unrolled with all the fanfare that Harvester could muster.

Not long after that, they would be introduced to the public at the 1981 Farm Progress Show. Farm Progress was the biggest consumer show for farm equipment in America, and it drew people from around the country and the globe to see what was new.

In modern times, the big show is held at one of two permanent locations in Iowa and Illinois that were built specifically for the show. The major manufacturers have permanent buildings on site—it's an incredible display.

Back in the early days of the show, it was held on working farms. In 1981, it was held on Schachtrup Farms near Brimfield, Illinois. Mike Schachtrup runs the farm with his brother, Steve. They actively sought out hosting the show.

"In the seventies, it was a rotating show—it would go from Iowa, Illinois to Indiana. I knew they would be coming to central Illinois in 1981, so I contacted them in '79. They came out, we had some discussions, and I got along with the hierarchy. They agreed on having it here in '81."

Schachtrup recalled how tough farming was in the day, with interest rates rising and land values plummeting.

"Every day you'd wake up and you'd be worth $10,000 less than the day before. During those years, Steve and I basically worked for nothing. If anything was wrong, you fixed it yourself—you couldn't afford to pay anybody."

He believes hosting the show helped his family survive.

"So then we had the Progress Show. With Dad's ability to wheel and deal with all these vendors, it was right up his alley. He'd say, you want to have signs up here on this parcel, you want to have this or that showcased, well here's what it's going to cost you. [Laughs.]

> "Every day you'd wake up and you'd be worth $10,000 less than the day before. During those years, Steve and I basically worked for nothing."
> – Mike Schachtrup

"We got three years of free seed out of it, and a year and a half's use of machinery. It was a year and a half to set up, and over a year to tear it down. It was good for the community. The churches made a lot of money; they got all the concession sales."[96]

Schachtrup Farms and the 1981 Farm Progress Show would provide a showcase for the Harvester's new 50 Series—as well as a chance for Youle's salesmanship to take center stage.

These people are part of the 50 Series launch crew, shown unwinding at the end of the day. *Roger Lewno Collection*

BLAZE OF GLORY

Fall 1981

Kansas City, Missouri

As the engineering team was putting the final touches on the new-for-1981 tractors, Youle had been battling to get farmers in red equipment and fighting for the products he felt the company needed to survive.

He argued with his boss, Stan Lancaster, about the need for a powershift transmission on their tractors. "Lancaster and I . . . fought over this for three years," Youle said. "I couldn't put a powershift in it."[97]

Whether Lancaster wanted one or didn't was immaterial—Harvester didn't have enough money to develop and manufacture the new technology. Deere's powershift transmission had been out nearly two decades, and the green machines were killing the red in markets where the technology was key.

"My friend with the Yellow Wheels had the powershift," Youle lamented. "The [John Deere] 4840 was running wild in the Mississippi Delta!"[98]

The synchronized transmission in the new 50 Series tractors was presented as an antidote. Youle, busy pushing Snoopy sales and cleaning

The new 50 Series and 30 Series Harvester tractors were introduced in Kansas City in fall 1981. The 2+2 tractors also received a new moniker, the 3588 becoming the 6588 and so on. This was largely a branding change, as the engine and drivetrain were not significantly different. *Wisconsin Historical Society / 111006*

up the Mexican tractor mess, recalls being told to stay away from the launch of that machine. That is, until the launch went sideways.

"[The] first show was at nine o'clock in the morning. They fired the whole bunch at nine thirty."
– Bud Youle

Harvester had hired an outside agency to handle the launch of the 50 Series tractors. The agency hired a stand-up comedian from Mississippi, Howard Gerald "Jerry" Clower, to emcee the launch. Clower had a number of hit records that consisted of him telling entertaining stories about his country life. He was wildly popular, and his earthy, G-rated humor seemed to be a good fit for a farm machine launch.

The problem was the emcee of the show didn't really understand tractors—particularly their transmission. In fairness, it was a fairly tricky thing to explain. The transmission was not a powershift—but you could do some shifting on the fly.

No matter, the emcee reportedly made a joke that you had to be "left-handed" to properly operate the tractor's new transmission. This was a gaffe of epic proportions—the new gearbox was perhaps the most important feature on the new model.

"It was horrible," Youle said.

The joke went over like a lead balloon with the Harvester executive team.

"[The] first show was at nine o'clock in the morning," Youle said. "They fired the whole bunch at nine thirty."[99]

According to Youle, the reaction at Harvester was to call in the cavalry.

The call went out, across the company—where in the hell was Bud Youle?

Youle was soldiering away at the Farmall Plant.

As Youle tells it, they called Matt Glowgoski, the head of the Farmall Plant.[100]

"Matt, get him down here—helicopter, airplane, boat. . . . Get him here!"

Helicopters or no, Youle was delivered to the Schachtrup Farm, a red machine prophet sent down to preach to the Farm Progress Show faithful.

This presentation took place at the 50 Series launch held in Kansas City in 1981. *Roger Lewno Collection*

When he met the group, he was handed a thick book packed with technical specifications for the new tractors.

"You got forty-five minutes to memorize it," Youle recalled being told.[101]

The new tractor's presentation was complex. There were instructions on what to say and when to say it. There were mechanical curtains and doors on the stage that would open and display material to explain features and technology. The massive volume was too much to memorize in a day, much less forty-five minutes. Not to mention coloring within the lines wasn't Youle's style.

"I said to myself, I'm going to do it my way.

"I walked out there with 4,500 people sitting in that audience . . . and [I] did what the marine corps told me to do—don't stop, keep moving."

He opened with a colorful line.

"We're going to go through those Yellow Wheel people like crap through a goose."[102]

Boom. The crowd roared. He was rolling.

"I had the ball, and I went through that whole tractor. I didn't need one thing that was in that book," Youle said. "I made a name for myself that day."

The crowd loved him. One can only imagine the reaction of the poor schlub who wrote that thick book of ignored instructions.

The fired comedian, Jerry Clower, lost the emcee gig but was still at the show. He and Youle had a grand old time.

"We had ladies from that state of [Illinois] bring in every kind of pie you could think of because Jerry loved pie. We had pie like you can't believe."

"But I'm going to tell you, we got the tractor introduced."[103]

Youle's performance on stage was not by the book, but it was a hit.

"We're going to go through those Yellow Wheel people like crap through a goose."
- Bud Youle

The "yellow wheels" reference to Deere became his signature bit, and Youle started using it every chance he found.

"I wound up having to go all over the world, damn near doing the same show over and over again," he said, adding that he spotted people from Deere in the crowd at other exhibitions, taking photos of him. "If somebody come up to me and ask me a question from Yellow Wheels or from Massey, I'd tell 'em. I don't have anything to hide," Youle said.

Times were tough, both on the farm and for the company. The farm crisis continued to create havoc, with interest rates up and land values and crop prices down. Farmers were struggling to survive, much less buy new equipment.

Harvester's fiscal situation was dire—it was shedding salable properties to raise cash, fighting with bankers, and saddled with debt, lawsuits, and a skeptical public.

Among all this turmoil, Harvester's sales machine was running at full throttle.

"We're making iron right now. Farmalls, they're rolling down the line," Youle said. "We all fought right here in 1981."

That year, Harvester recorded near record-breaking sales. The best year for agricultural equipment sales in the company's history was 1979, with $3.07 billion. The Harvester crew chalked up $2.98 billion in ag equipment sales in 1981.

"We thought, this is it. We're going to make it," Youle said. "We're going to make it."[104]

Jerry Clower was a comedian who pulled from incidents in his life growing up in rural Liberty, Mississippi, to entertain. A former seed salesman, he got his start telling funny stories to customers. One of his first major appearances was with Charley Pride, who took him under his wing. *Author's Collection*

CLASH OF THE TITANS

1981

The Heartland

Competition in the tractor market is a game of cat and mouse that has been played for more than a century. In the 1970s and 1980s, the game reached new heights between Deere and Harvester.

Kelly Kravig was a marketing specialist for Steiger and then for Case IH. He explained how the game was played in that era: Deere would wait for Harvester to introduce a new model or update and would follow that up with its next machine—and it made sure that machine had enough extra horsepower or features to have a bit of an advantage on Harvester.

Note that sales were garnered by comparative marketing, which basically means showing that your line has better features and more horsepower than the comparable models offered by the competition. More horsepower has been the key sales point for tractors since the 1950s. While this was particularly true in the 1960s to 1980s, the basic precept carries into modern times.

The key to playing this game well was to be constantly investing in research and development, ensuring that whatever the competition did, your company had new technology tested and ready to move into

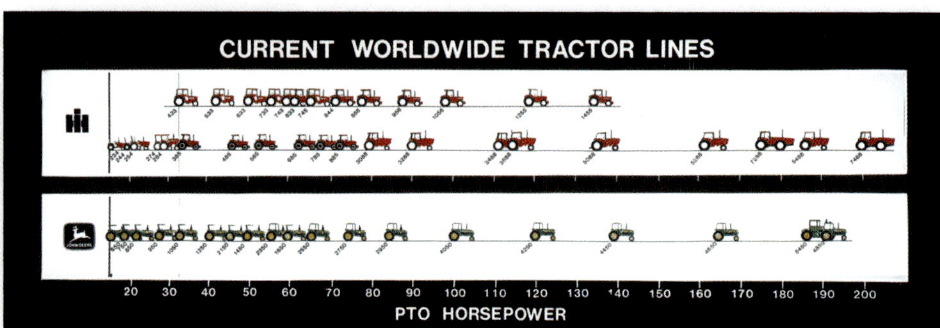

This graph was created by Harvester to illustrate the market positioning it planned with its new machines. Kelly Kravig noted that colorful illustrations like this became popular ways to show engineering and marketing plans to executives, who reacted more positively to graphics than to spreadsheets. *Adam Summerlott Collection*

production when the competition showed its hand. You can see how this played out through the 1960s to the 1980s. The Deere model line launches closely followed the Harvester model line introductions—and consistently offered more horsepower.

Harvester was able to break this cycle with two key products in the 1970s: the Axial-Flow combine and the 2+2 (Snoopy) tractors. The Axial-Flow (and New Holland's twin-rotor TR-70) brought a new technology to market. This gave Harvester an edge it could exploit until the patent ran out on the Axial-Flow. Harvester also had an edge with the 2+2, which filled a market niche in which Deere did not have a competitive entry.

> "Deere model line launches closely followed Harvester . . . and consistently offered more horsepower."

An updated launch of combines exemplified that constant jockeying for position. Case IH was planning on updating its combine models, but it also knew that Deere was launching a new combine series with an all-new cab. Case IH delayed its product launch until after Deere released its new machines. Case IH then continued its focus groups to get feedback from customers on the new Deere cab and the proposed Axial-Flow cab updates. It learned that customers liked the fridge option in the Deere, but it discovered that customers preferred a portable fridge over one that was fixed in the combine cab.

The net result was that when the updated Axial-Flow was introduced it had a quieter cab, more usable space, better ergonomics, and a portable fridge just like the customers wanted. That was the game.

In 1981, the big issue for Harvester was that it had been playing catch up since the mid-1960s. More critically, Deere invested massively in research and development throughout that period. This meant whenever Harvester rolled out an improvement, Deere could respond and release its new upgrade quickly.

DREAMING OF DADDY WARBUCKS

1981

Chicago Suburbs

Gerry Salzman was a young up-and-coming marketing executive for Harvester in the early 1980s. A farm kid from Ashton, Illinois, he had grown up using red equipment. He started his career at Harvester as a territory salesman and was quickly moved to Chicago and made the marketing manager for harvesting equipment.

Like Youle, he had a gift for making connections with just about anyone.

He and Youle had offices near each other. "Bud was right across the partition from me," Salzman said. "It was hard to get stuff done because he was so loud!"

Salzman had been in Kansas City and at the Farm Progress Show for the launch of the new product in fall 1981. "The 50 Series was supposed to get us back into it again."[105]

The equipment was good-looking, and even without all the technology everyone hoped for, it was a leap above and beyond the previous machines. The problem was the times. The entire economy was down, and farmers were suffering.

At the end of 1981, Harvester had recorded a loss of $635 billion. It had not made a loan payment since March. McCardell's magical money tree had finally ran out of fruit.[106]

As Harvester's fortunes crumbled, so did those of the farmer. This is a farmer and his grandson at a 1986 livestock auction in Siler City, North Carolina. *Rob Amberg*

"When is your &#@!ing company going to pay our bank?"
— Disgruntled Banker

Salzman was commuting to work at the Harvester offices downtown. "I took the commuter train down to the office and ran into a fraternity brother who was in the banking business."

Rather than exchanging the expected banter about old memories from college, the man greeted Salzman cooly. He finally said what was on his mind.

"When is your &#@!ing company going to pay our bank?"[107]

When your company skips loan payments for the better part of a year, the blowback can reach all the way to your morning commute.

Money was one of Harvester's big problems, and the introduction of the new tractors that fall did not help matters. The big news in December 1981 was that the company was able to negotiate a new $4.15 billion financing deal to stave off bankruptcy.

As an aside, Glenn Kahle had said that executive Warren Hayford, in particular, was an advocate of bankruptcy. "Warren Hayford, he pushed like everything to take it into bankruptcy. In fact, we went to the Bahamas one winter over Christmas and I had a packet of the bankruptcy thing so it was going to be announced while we were down there, and it turned out that some bank down in Mississippi changed their mind and they got taken over or something so it didn't happen."[108]

Farmers of the 1980s had a lot to consider. These farmers are talking with farm organizer Charley Thompson at a feed store in Harmony, North Carolina, in 1986. *Rob Amberg*

Kahle was no fan of Hayford's, and he wasn't alone. Stanley Lancaster took early retirement in 1981, and he pulled no punches about why he

“I said I’m not coming back for some Mickey Mouse job reporting to some young MBA.”
- Stanley Lancaster

left. McCardell’s “biggest mistake by far was the infusion of people not familiar with our business and what makes us tick.” He later came back to Harvester, and said, “I said I’m not coming back for some Mickey Mouse job reporting to some young MBA. I said I’d have to run the whole thing.”[109]

The loan gave Harvester a lifeline to the future. At the time, analysts expressed skepticism about that future. “The pact doesn’t solve any problems except immediate bankruptcy,” one said.

William Roberts of Piper, Jaffray, Hopwood Inc., a widely quoted agribusiness analyst, was blunt. “The bankers should hold up two fingers in front of their faces and have them say, ‘We’re number two,’” he said. “There’s no way, even if they had a Mideast Daddy Warbucks peeling it off, that they’d be number one.”

McCardell appeared in a press conference in shirtsleeves, swore they wouldn’t sell off any of the core divisions, stating there was several hundred million dollars’ worth of noncentral assets that could be sold, including a “piece of property in Arizona.” The property was likely the test grounds near Casa Grande.

Harvester engineer Steve Warner was young when this was going on, and he remembers the days vividly. “In those days I was pretty naïve. . . . I was just a young engineer coming into work and didn’t really give a lot of that a whole lot of thought. We had regular visits by the banks when they would do the refinancing.

Warren Hayford (left) was hired by Archie McCardell to help with cost-cutting measures. This did not make him the most popular executive. He is shown here touring the Harvester plant in Fort Wayne, Indiana. *Wisconsin Historical Society / 144399*

That was really strange to see three or four buses pull in out front of Burr Ridge and gobs of bankers get out and come through the facility every, it seemed like it was every three or four months there for a while."

Mike Bollin went to court to try and gain the legal right to remain on his farm after foreclosure. He and his wife, Kathy, fought for their rights and his efforts helped draw attention to the plight of farmers in the 1980s farm crisis. *State Historical Society of Iowa*

Kahle remembers having to keep the bankers at bay: "The bankers had supported us all through this and we had a bank meeting at least once a month with probably, oh, I'm guessing now, ten, fifteen, maybe twenty, sometimes, more bankers would show up and we'd put on what we'd call the dog and pony show, which was really what was going on and show them the progress we were making. And so we kept our engineering funding up way beyond what you would have thought of for a company in [as] bad a shape as it was."

Youle understood that for his company to walk out of this, they would have to get the cash. They would also have to keep the fight in the fields. "We either had to have the money, or we were going to put a for sale sign out in front of the building on 401 North Michigan [Avenue]. That was it, because we knew that Deere was coming with the 4850. They were pushing this thing up. They had updated their 466 engine."

Competition was out there that could take away the little market edge Harvester had with Youle's beloved 2+2. Deere was working hard on a much-improved mechanical front-wheel-drive (MFD) tractor—and much more.

"We knew, we could sit and see where they were going. We knew they were going to come with an MFD. We didn't know when."

Youle would soon find out.

THE FURROW
FIGHTING RURAL CRIME
CANADIAN PEANUTS TAKE OFF
WHEAT TILLERING PINPOINTS STRESS
The commitment is forever.
DEPUTY
GUIDED TOURS OF THE SUPERDOME
DAILY
Every Hour on the Hour
9:00 A.M. to 4:00 P.M.
Tours Include
Take the Superdome Tour and HAVE LUNCH IN THE BEAUTIFUL STADIUM CLUB
FOR RATES AND DAILY TOUR INFORMATION CALL 587-3810
We're going on.
IH FARM FORUM
CANADIAN EDITION
PLANNING 1982
GUARANTEED BUSHELS
THE LAND RENT SYSTEM OF THE '80s?
SUPERSTAR
Maison Bourbon
JAZZ
Now you can buy that new John Deere combine and forget about finance charges until 1983
Bayou Macon Equip. Co.

Chapter Five

THE SUPERDOME SHOWDOWN

1982

EVERY LITTLE BIT HELPS

January 1982

Detroit, Michigan

On January 14, 1982, the UAW and Harvester officials came out of a closed-door meeting in Detroit at an impasse. Harvester had asked for $100 million in wage concessions in November 1981, and the UAW had once again refused to budge, citing the "outrageously high salaries" of Harvester executives.[110]

The highest-paid of those, Chief Executive Officer Archie McCardell, had overseen a $1.1 billion loss for Harvester over the previous two years and had a large percentage of its union labor workforce laid off following Christmas of 1981.

Complex banking loan restructures were slowly unfolding, which were critical to the company's solvency. Bankruptcy loomed on the horizon, and the union's fortunes were tied to the success of the banking negotiations.

"Who wants to buy a truck or a tractor from a bankrupt company?" a union official said. "We're praying the bank deal works."[111]

As the cold winter turned to a grim spring, many of the much-lauded executives recruited to save Harvester—and paid handsomely despite their failures—pocketed their last fat paychecks and headed out the door.

In March 1982, Warren J. Hayford abruptly resigned. McCardell publicly expressed regret—not surprisingly, as Hayford was his hire and his right-hand man. Little else was offered from Harvester on the departure, which was widely speculated to be a move to appease the unions, the bankers, or both.

His high salary, battles with organized labor, and Harvester's lack of success left McCardell as the whipping boy of the times. In July 1982, *The Boston Globe* ran a photo of him captioned, "The wrong man at the wrong time." *Wisconsin Historical Society / 23051*

As the 1980s wore on, Harvester continued to make tough cuts. These men were laid off in September 1982, nearly two months prior to the photo being taken. *Chicago Sun-Times*

The Washington Post would later report that Hayford left on his own accord because McCardell wouldn't listen to his pleas for deeper cost cutting.[112]

Whatever Hayford's final play, neither the bankers nor the union was happy with Harvester that year. The UAW and Harvester came to terms in spring 1982. Perhaps not coincidentally, Archie McCardell was out a few days later.

He resigned at the request of the company directors. A company spokesman said, "The company obviously has had heavy losses for the past few quarters. The board felt the financial situation required a change."[113]

While Harvester and UAW officials denied any link between the agreement and McCardell's departure, speculation that the two events were tied remained. "There's no doubt that there's been bad blood between the unions and McCardell since the strike," said Eli S. Lustgarten, an analyst with Paine Webber. "Those of us who have been watching the situation have been doubting McCardell would survive."[114]

Harvester had a replacement lined up when the McCardell announcement came, indicating the move had been planned for some time. The new head of Harvester would be Louis W. Menk, the former head of Burlington Northern Railroad. Known for cost-cutting measures—he was the man who killed some of the railway's most beloved and least profitable lines—Menk was described as a "perpetual motion machine" as well as a "snake charmer."[115]

Menk had been coaxed out of retirement to take over Harvester. Known to be an engaging person and a careful manager, the word on the street was that he was put in place to make the tough cuts Harvester needed, manage the endless

This is an equipment auction at the IH dealership in Marshaltown, Iowa, which was going out of business. *New York Times*

morass of negotiations with bankers, and carry Harvester through the nearly inevitable sale or bankruptcy that appeared to lie ahead.

Menk was joined in the leadership ranks by Donald D. Lennox (the former head of Harvester's manufacturing group) and William G. Karnes, who was brought in from Beatrice Foods Co. to chair Harvester's finance committee. The leadership team brought a lot of age and experience with finance to the table, which led to more speculation that this team was there to manage Harvester's exit.[116]

Gerry Salzman remembered that around the time of the transition his phone would ring on a regular basis. All too often, the calls would start with, "'Hey, I heard a rumor.'"

Salzman soon tired of this practice. He wasn't in the loop for that kind of thing, and he had work to do selling equipment.

One day, he got fed up with another caller asking about sales rumors. "He was telling me some bullshit. It was always, 'Someone is going to buy us.' I told him we were going to be bought by Walt Disney. 'Yeah, you are going to work with Donald Duck and Mickey Mouse. Leave me alone!'"

Salzman added that working for Harvester at that time was a challenge. "It was a love-hate relationship. You wanted to stay and nurture your career, but you knew the company was in big trouble and we didn't know what was going to happen."[117]

> "When you're going to lose $700 million, like Harvester will this year, every little bit helps."
>
> – Eli S. Lustgaren

As 1982 slid into the worst El Niño weather pattern in recorded history—one that brought dust storms, droughts, and brush fires around the globe, not to mention an unprecedented outbreak of rattlesnake bites in Montana—Harvester weathered its storm by selling everything that wasn't bolted down.[118]

It sold its mine holdings in late June and five parts centers a few days later. It put a nodular iron foundry in Waukesha, Wisconsin, on the block on July 2, 1982, prompting analyst Eli S. Lustgarten to comment that the sale "is no big deal, but when you're going to lose $700 million, like Harvester will this year, every little bit helps."[119]

MISSION IMPOSSIBLE

July 1982

New Orleans, Louisiana

While the predictions of the time were dire, the folks committed to the tractor industry—for better or for worse—continued to work at making a difference. Stanley Lancaster, Bud Youle's boss, became frustrated with leadership and exited. His replacement would prove to be one of the bright spots in the grim months of 1982.

Robert Carlson was the man tapped to replace Lancaster. Affectionately known as "Tractor Bob," Carlson came to Harvester from Deere. Youle liked him—and so did pretty much everyone else.

The Superdome in New Orleans, shown in January 1983. *Library of Congress*

"Bob Carlson was now the president of the Ag Division," Youle said. "Bob Carlson was the type of person that came from the Yellow Wheels. I could look Bob Carlson in the eye and say, 'This is what I want to do. This is why I want to do it. This is how I want to do it.' He said 'You're wasting time talking to me.' He said 'Go'—that's the kind of person he was."[120]

Youle had found out that Deere was releasing its new line of tractors—the 50 Series. The machines were being manufactured at a brand-new plant in Waterloo, Iowa, and they would have a much-improved mechanical front-wheel-drive system.

He wanted to know how they would compete with the Harvester machinery, and he had a lead on where they were being tested. He asked Tractor Bob about that.

"I said, 'I would like to go to Laredo. I know that's where their test tractors are running,'" Youle recalls saying, adding that he knew Deere had a test farm down there.

"'Get out of this office and go.' That's the way Bob Carlson was."

So, Youle went to Laredo, Texas, to see what he could find. He knew the tractors were being tested somewhere in the area—he just didn't know precisely where that might be.

He took off his signature Harvester sales guy garb and got into the uniform of a working farmer. "I didn't dress like this. No, no, no, no. I had a little grease on me here, a little grease on me there. Walked into [a] John Deere dealership. I said, 'I need to buy an oil filter.'"

Sitting outside the dealership, Youle spotted a vehicle outside with "John Deere Test Engineering" lettering on the side.

"That was my vehicle. I didn't know what farm they were working on. So, I go out and get my rental car [and] followed this man up over hill and dale."

Sure enough, Youle found the machines out in the field. And, as he suspected, Deere had made engineering changes to the front end of the mechanical front-wheel-drive system. "Big surprise," he said. "They got the tires laid over to the side when they turned—made 'em turn easier and they turned shorter."[121]

One of the major problems with front-wheel-assisted tractors prior to the early 1980s was that they took forever to turn around. The heavy drive systems and big front tires made the turning radius a country mile. Some engineers figured out that laying the front wheels to the side would let the big machine turn around in a smaller space. Deere had this feature on the 50 Series in 1982, and Youle spotted it.

"I didn't know what farm they were working on. So, I go out and get my rental car [and] followed this man up over hill and dale."

- Bud Youle

"They leaned the tire over, because they couldn't get the shortest turning radius that they wanted to get. That's how they made it get closer to what the 2+2 was, because I'm convinced until somebody tells me differently that the Deere company was scrambling like hell to make a 2+2 tractor, because they knew the merits of that tractor."[122]

He suspected the new Deere wouldn't turn tightly, even with the laid-down wheels—but he needed to gather more information to be positive.

Youle proposed a trip to Tractor Bob that had two legs. First, he would head to Laredo to scout Deere's proving grounds. Then he would go to New Orleans. Youle had gotten wind that Deere would be introducing its new machine down there, and he wanted to get a look at it.

Deere planned to launch its new line in the New Orleans Superdome. Rumor had it that the show would be nearly as flashy as Deere Day in Dallas, where Deere had introduced the New Generation tractors in 1960.

John Deere territory manager Dale Johnson remembers the day in New Orleans well. "Deere had rented the New Orleans Superdome for the new tractor intro, because it was going to be a huge intro with the new 50 Series tractors with the fifteen-speed powershift transmission and the mechanical front-wheel drive and more horsepower," Johnson recalled. "They had laser lights and live music, and they had dancers and performers, and it was smoke and fire."[123]

With dealers from around the world, the event was quite a show.

When Youle arrived in New Orleans on that fateful hot day in late July 1982, the Deere team had machinery outside the Superdome, with a horde of young guys waxing them in preparation for the show. Youle went dressed incognito. "When I come walking up, I had an old shirt on, pair of pants. I didn't have any Harvester identification," he said.

Youle was flying below the radar—just another dealer scouting his sales prospects for Deere's new machine in the down disaster of a market that defined 1982.

Youle wanted the 4-1-1 on the new rig. He eyeballed it up good, then checked his eyes by drilling the natives.

"Is that the 466?" Youle would ask. Casual like.

He was referring to the engine. Deere's 466 was a venerable unit. Nothing new. Nothing to sneeze at.

"Oh yeah, that's the 466."

"That's the mill. How's it tuned? What are you running that baby at? 180?"

Engines can be tuned. Factories play games, tuning engines below capability. Customers blow fewer of them up that way.

Deere's new line of 50 Series tractors were built in a brand-new, state-of-the-art facility in Waterloo, Iowa. The plant featured computer-controlled assembly that was said to cut inventory and reduce parts shortages. This is the plant in April 1988. *Author's Collection*

Youle knew the 466 was good for 200 horses. Had tuned one himself. Figured Deere would play it cool.

They did.

"Oh yeah, about the same way. Set it up about the same way as on a 4830."

Youle kept walking.

"Now, I didn't get inside, because they had a guard standing there at the gate. I didn't want to create any turmoil about going inside and seeing their arena show."

Youle knew going inside was a bad move. Outside the unwritten rules.

But his company was in trouble; his beloved 2+2 had a challenger, and he was Bud Youle.

"When I come walking up, I had an old shirt on, pair of pants. I didn't have any Harvester identification."
– Bud Youle

"I went back to the hotel, and I got to thinking and figuring out.

"You know, I could put a Deere hat on and a shirt.

"No, I don't want to do that. I don't want to do that.

"If I did that, that'd create an international incident.

Youle reputedly dealt with the police. It may also have just been Superdome security. This is a New Orleans police car in 1976. *Library of Congress*

"I'm not going to do that."

He had been in the Astrodome before.

He knew there was a bar up top where he could hear the sales pitch. Size up the new competition. Maybe toss back a longneck or two.

"That's all I needed to do. I didn't need to see the guy. I wanted to listen."[124]

He decided to do that.

In the morning, he walked the Superdome grounds. The kids with the wax and cloth were gone, so were the new green tractors.

"I walked in the front door of the New Orleans Super Dome just like I owned the place."

Just like that, he was in.

Tom Renner was one of the hundreds of John Deere dealers inside the Superdome that day. Renner runs one of the few remaining independent John Deere dealerships, the Shiloh Valley Equipment Company in Belleville, Illinois. Unlike the big dealership conglomerates that run dozens of locations, his dealership is just one location. His family has worked a farm in the area since 1847. He's the seventh generation to work the land and has family that works the farm and his dealership.[125]

Renner founded the company in 1969 and started selling John Deere equipment in 1972. He said that new equipment shows for dealerships in the 1970s and 1980s were much grander affairs than those at present. He remembered the 1973 edition with particular fondness, admitting that might be because it was his first one (as Renner recalled, Tammy Wynette played).

Deere introduced its new Generation II tractors, the 30 Series, at the show in Waterloo, Iowa. He was blown away by Deere's new planter, which was more capable and expensive than anything that had been on the market. He bought several and recalled selling them well to his customers, but he said that pricing was destroying sales in the early 1980s. "Price tags got too high. The big four-wheel drives were selling for $130,000 to $140,000. That was a lot of money."

"If I did that, that'd create an international incident."
– Bud Youle

Even in those dark times, however, customers would buy equipment—if the price was a bargain. There were plenty of bargains to be found. Banks were regularly repossessing farms, and they would sell the equipment off cheap to recover some of the loss.

The manufacturers were also dumping equipment. As Deere was winding down the 40 Series, Renner remembers picking up new machines for nearly half price. He could sell those machines, but Deere added a kicker to the deal: you had to order a minimum of two to get the model into your dealership.

"If you sold one, you had to buy two," Renner said. "A lot of guys hesitated—why would I want to double my volume?"

As far as the 1982 launch of the 50 Series in the Superdome, Renner didn't recall being overly impressed by the event. He went to Bourbon Street; he didn't care for that scene and got sick when he ate some bad shrimp.

He also said that you had to present a badge to get into most of the dealer equipment launches back then—these passes were typically included in a packet that dealers received when they checked into their hotels. He also recalled that while the badge was required for admittance, the place was not locked down particularly tightly.

"Security-wise, I know I could sneak in."

Dale Johnson also remembered that the event required badges to get in.

That didn't stop Bud Youle.

He executed his plan. Took the elevator straight up to the top.

"I walked over. I sat down up there in the dark. Where I was sitting I was looking right down on the arena presentation."

He thought to himself how pissed off his people would be if a Deere guy snuck in.

Then he listened.

The words he overheard that day were familiar to him. They were talking about his beloved 2+2. "I kept hearing terms like, 'They've got this Snoopy tractor that can do this. They can bring it into thirty-inch rows.'"

"Really, I sit there. I said, 'Did I write this damn script for them?'"

One of the key things Youle learned that day was the turning radius for the new Deere was not tight. In the words of a chat board denizen, "Seems like it takes a barnyard to turn it around."[126]

Deere would develop the technology to make the machines turn more quickly—on the next generation of machines. For the time being, Youle had found an advantage for his machinery.

But he was in enemy territory. Enjoying a beer or no, he was surrounded.

"I'm sitting there. The marine corps taught me one thing that I failed to do. You never stay in one place too long. You move."

"I thought, 'If I move up to that bar, there's a light up there, somebody might be looking around. I better stay in the dark here.' I got myself a Budweiser beer and walked over and got in a different place, but I could still hear them."

In the meantime, the green tractor people realized they had a mole.

"During the show, one of our security people thought he saw some kind of a reflection on the roof of the Superdome," Dale Johnson said. "They've got a huge structure up there and catwalk with lights and speakers and they hang the big screen from there."[127]

Johnson remembers security telling him he was headed up to investigate.

The reflection on the Superdome's roof was Youle.

He thought to himself how pissed off his people would be if a Deere guy snuck in.

He saw the writing on the wall.

"I could hear them," Youle said. "Pretty soon, I heard the elevator come up. I remember that. Two guys got off the elevator."

A security guard approached Youle.

"My name is John Doe, Vice President Security, John Deere Company, East Moline, Illinois."

Youle said there was another guy, who said,

"My name is so-and-so. I work for him."[128]

Youle was escorted out of the Superdome. This is not Youle. *Library of Congress*

Johnson remembered his security guard reporting back. "He went up there and low and behold, he came upon this guy that was taking pictures," Johnson said. "He called down for some help and then ended up grabbing a hold of him, so they got him down off the catwalk and got him into their security room and then started asking him a whole bunch of questions."

Youle recalled the security guard was a big man.

"He must have been six foot six inches . . . weighed 285 pounds," he said.[129]

"The minute they arrived, I handed them a card. I said, 'My name is Bud Youle, Harvester company. I was down here, and I saw what was going on.'

'We're here to throw you out and put you in jail.'"

Youle remembers pointing out that the Superdome was public property (it's not). He says they then changed their tune (they might have).

"'Would you mind leaving Mr. Youle?'

'Yes sir, I'll leave.'"[130]

"He was going to put handcuffs on me. I said, 'No, you're not going to put those on me. No, no, no, no, no.'" [131]

"I had a piece of paper in my hand. I said, 'Here, you might as well have this piece of paper, because all I did was confirm the transmissions and the turning radius of the tractor.'"[132]

He also recalled the Deere guys were a bit upset.

"Of course, the guy there, I remember he got a little huffy with me."

Youle claimed to have reminded them how he caught Deere sneaking around.

"You remember when we started shipping 2+2s in those boxes? Right above the tractor plant at Rock Island up on top of that hill was a girl's catholic school. When the first tractor rolled by that building, there must have been forty-nine cameras taking pictures of that box pulling that railroad, pulling those box cars out of there."

I said, "Do you remember that?"

The guy stood there.

"This is an espionage business. I caught you. You caught me."

He also said they just tossed him out and didn't arrest him. He said the big cop (aka former football player) was a friendly fellow who gave him a ride back to his hotel.

"Here I am, five foot nine-and-a-half inches, 175 pounds ringing wet. He could have picked me up and down the elevator we'd go. I asked him, 'Have you ever been to Missouri?' I got to kidding him, because I was positive they told him he's not going to take a police car to take me downtown and put me in jail and book me or anything like that.

"I said, 'When I get downtown in the street here, you're going to put me in the backseat, you're not going to put handcuffs on me, are you?'

"'No. No, I'm not going to do that.'

> "He was going to put handcuffs on me. I said, 'No, you're not going to put those on me. No, no, no, no, no.'"
>
> – Bud Youle

"The policeman was very nice. I kidded him all the way down about what he had to do. I said, 'It's your job. You kicked me out of the building. I'll be out of here. I'll get my bags packed. I'll be out of town.'"[133]

As Johnson told the story, Youle was held for a bit in the Superdome before being sent home. Johnson recalled getting on the horn and making a series of phone calls, trying to figure out what to do with him.

"They had an idea of what they'd like to do with him," Johnson added with a laugh.

In March 1982, Youle spied on the Deere introduction from the top of the Superdome—likely above these seats that are pictured here. *Author's Collection*

They finally were able to reach an executive who helped them figure out how to properly dispose of the spy. Johnson and Youle both recalled the executive and the new CEO at Harvester, Louis Menk, having words to settle the matter.

According to Johnson, the executive called Menk and said, "We're trying to figure out what we should do with your guy here. We've caught him taking pictures and we don't know whether we should shoot him or just let him go back to you."

Menk must have suggested the latter option, as the Deere folks took away Youle's camera and released him to scamper home to Harvester.

Shortly after Youle got home, he took his knowledge to a national sales meeting, where he was able to let his fellow Harvester faithful know that they had a sales edge to use on the new hardware from the green guys.

"The point of it was I was able to sit with the company people and express to them how the tractor was going to turn, how the wheels were going to turn, how the wheelbase had been increased. That Deere had done everything they could to make a shorter turning radius tractor. They couldn't do it. They could not get it done."

"We've caught him taking pictures and we don't know whether we should shoot him or just let him go back to you."
– Deere Executive, in a phone call to Louis Menk

Youle recalled telling them, "We don't have to take a backseat. The 5288, the 54 will take care of itself. We can turn shorter.

"We've got the 72, 74 coming on the 2+2s. We're going to make fools out of them."

Youle was referring to the existing 50 Series, which turned shorter, and a line of new high-horsepower 2+2 models planned for release in 1984.

The engineering team would eventually discover another flaw in the new Deere.

"On the MFD you got to have a lead and a lag. We were shooting for about a . . . in fact, the 50 Series, I think it was 1.5 percent lead, which meant the front tires turned one and a half times faster than the rear wheels. Deere could not get there without a major retooling of the power train. They had a minus one or something like that. I remember that."

The new Deere 50 Series were introduced to dealers in an invite-only event held in the Superdome in New Orleans. The MFD models presented a potential threat to IH's 2+2 line. *John Deere*

Four-wheel drive requires the front and rear wheels to run at the same speed. This is fairly easy if you have equal-sized tires, as the 2+2s and the big four-wheel-drive tractors from most manufacturers did at the time.

Front-wheel-assist tractors present more engineering challenges, one of the trickiest of which is getting the speed between the front and rear wheels to match. With a rear-wheel-drive machine, this is not an issue. Drive the rear wheels and let the front spin freely.

When you got to a front-wheel-assist system, you have smaller tires up front and bigger tires in the back. That means you have to carefully calculate how fast to turn the front wheels so that the speeds match perfectly with the rear.

Bear in mind that this changes if the owner uses different sized tires for the front and rear.

All the manufacturers struggled with this issue and, as Youle noted, Deere couldn't change things enough on the 4850 to match speeds perfectly.

This meant the new model was going to wear tires quickly—this was valuable intel for the red marketing staff. Youle used it often when he dealt with the Yellow Wheels folks.

"I always remember to be sure to tell the Deere people that they're going to have trouble wearing out tires," Youle said, adding that his colleagues at Harvester did the same. "The guys had a lot of fun with this."

Harvester had a brand-new CEO in 1982, one Louis Menk. He came from the railroad industry and was known for his cost-cutting measures. He also was dragged into the Youle spy incident. Given he was trying to lead a company into a sale, bankruptcy, or worse, one can easily surmise he was not smiling like this when he learned of Youle's antics. *Wisconsin Historical Society / 36411*

THE FALLOUT

August 1982

Chicago, Illinois

The people who didn't have fun with all this were the leaders of Deere and Harvester. Youle's shenanigans were unprecedented. While it was accepted that you could peak under hoods when machines were found out in the field, sneaking into a corporate function was just not done.

Both leaders had their hands full at the time. Deere was in better financial shape than Harvester, but it was confronted with the same horrific marketplace. It was facing layoffs and had to scramble to build whatever it could to survive.

And the newly appointed Menk was up to his ears in bankers and litigation and a failing company in the worst economy of modern times.

Neither had to be terribly keen on spending time mending fences because of corporate espionage. Gerry Salzman remembered that Youle's job was at risk. "Bud was lucky he still had a job after that—I heard that."[134]

He may have kept his job, but he certainly caused a stir. The word at Harvester was that there was a meeting in Chicago, and Menk and Youle were there along with unnamed executives from Deere.

"Naturally, I had to go back to Chicago. When I went back to Chicago, I had a phone call from Mr. Menk. Mr. Menk asked me to come downtown to the office, so I did. I went down there," Youle said.

"Bud was lucky he still had a job after that—I heard that."
- Gerry Salzman

"There was another fellow there . . . he came to the Harvester company from John Deere. He was president or executive vice president of the Deere company. He was in the meeting also.

"I told Mr. Menk. I said, 'Mr. Menk, we needed this information.' I said, 'I got our people all geared up and ready to go.'

This ad illustrates how difficult the times became for Harvester in the 1980s. *Author's Collection*

"I said, 'Mr. Menk, if we have to settle for some reason, that's fine. Tell them we sent them a bill for some front axles that didn't work very well.'

"The guy from Deere, I remember, he laughed.

"Mr. Menk said, 'This thing is over with. That's it. You caught us. We caught you. That's the end of it.'

"I left, because he said, 'I'll handle this.'

"I said, 'Mr. Menk, fine. I'll go downstairs. If you need me, call me back upstairs,' because he did all of it. Next thing I know, he came down to the eighteenth floor, because we were still on the eighteenth floor as I recall. He walked in there.

"He said, 'Bud, this whole thing is handled. Forget it. Let's go on about our business. Let's put them out of business.'

"I said, 'Fine, Mr. Menk. That's fine.'"

Epilogue

THE FINAL COUNTDOWN

The difficult events of 1981 and 1982 were not the last of the struggles for Deere, Harvester, or farmers. While the economic depression would peak in 1982, times would continue to be very tough for all involved.

Farmers would face a rash of bank foreclosures, driven by falling land values, sinking crop prices, skyrocketing interest rates, and nervous banks. Families would leave the farm in droves; some left by choice while others were tossed out by bankers and left with nothing.

This devastated thousands of farmers and drove the disturbingly common trend of bereft farmers committing suicide. Farm protests broke out across the globe, prompting Willie Nelson, Neil Young, and John Mellencamp to found Farm Aid in 1985.

Mike and Steve Schachtrup—who hosted the 1981 Farm Progress Show—were some of the farmers who made it through the tough times. To do so required careful management of resources and a survivable amount of debt, with a little luck thrown in for good measure.

"Now, do you think I would've been a mean hog in the bowl lot with that baby?"
– Bud Youle

"I think my brother and I have been successful because of our work ethic. We are like a yoke of oxen, pulling the plow in the same direction every day. And he's got different talents than I do. He's very talented with machinery; he can fix just about anything. I couldn't have done it without him, and maybe he couldn't have done it without me. My father always said that our biggest asset was each other.

"We could not have achieved such success without the hard work and sacrifice from previous generations," Mike said. "This in an occupation that you truly have to be excited about daily to go to work, or the small aggravations and failures would do you in."[135]

Deere would continue to grow and maintain its stranglehold on the number one position in the agricultural equipment market, at least as of 2024. Time would eliminate much of its competition, but the years weren't easy on anyone.

As for Harvester, as most know, its agricultural division was sold to Tenneco in 1984, which merged the group with J. I. Case, creating Case IH. The merger was difficult, and Bud Youle's beloved Snoopy tractor was one of the casualties of creating the new company.

The Harvester team had a new high-horsepower version of the machine ready to roll, but the plug was pulled on that program and the J. I. Case four-wheel-drive line was kept. Youle had three months left before his retirement, and he said he opted to keep his opinions to himself for a change.

This prototype for a higher-horsepower Super 70, the 7688, was photographed at Hindsdale. IH released only a handful of 7488 and 7288 models. While this one was never officially released, the prototype is in private hands. *Kenneth Updike Collection*

Deere continued to develop new equipment. This futuristic drawing is the imagining of Chuck Pelly, one of the Henry Dreyfuss Agency's star industrial designers. *Chuck Pelly Collection*

"Now, do you think I would've been a mean hog in the bowl lot with that baby?

"If the Lord had let me have this tractor right here, I tell you exactly what I was going to do. I was going to get that microphone out. . . . I was going to get in the seat of this tractor right here, and I was going to have every dealer in the United States and dealer salesman meet me in Cedar Rapids, Iowa, and we're going to drive that tractor right up to the front door of Waterloo, Iowa. I'm going to honk that horn and say, 'Come out and look at this son. I'm a bitch.'

"I guess the biggest regret that I had in my life . . . was the two plus two. I didn't have the nerve to look a man in the eye and say, 'You don't know a damn thing about farming.' He'd have fired me. I had to have three months to get my retirement. I was going to get my three months. That's the only regret that I have that I didn't fight hard enough for that tractor."[136]

Youle certainly did fight, however. One of the ironic things about the entire escapade of sneaking into the Superdome is the timing. Very shortly after the dealer show came to a close, Deere introduced its new tractors to the public at a show in Waterloo, Iowa.[137]

Long-time marketing executive Gerry Salzman remembered that the event was the talk of the company. "Everybody loved it. Talk about a morale booster, that was it.

"That was just Bud's profile. People loved to be around him; he had such enthusiasm."

Times on the farms wouldn't get better soon . . . but Case IH would take all the expensive research funded by McCardell's money tree and build the Magnum tractor. Dr. Glenn Kahle and a good portion of his fellow former Harvester engineers would build one of the best machines of modern times.

For a whole bunch of people who had red tractors in their being, the Magnum modeled an ideal: red tractors would live on. There was a lot of hard work to get past the sale in 1984—and, ultimately, much of the next decade—but they and their company carried on.

In business—just like in life and family—good times go, and hard times come. We all have opportunities to turn tail and run or to charge up that hill.

Harvester's company fortunes took turns that are hard to comprehend as you look at them in retrospect, but Youle, the US marine, never shied away from the fight.

"This was a great company; at one time we were destined to be a great company. We fell short, maybe because people like me that were spending money trying to get ahead," he said. "They can say this Missouri boy was a mule, but I'll tell you what, I had a battle plan laid out. I just didn't have the cash."[138]

The research and development done in the late 1970s by Harvester engineers in their Hinsdale research facility would be used to create the Magnum tractor in 1987. *Gregg Montgomery*

ACKNOWLEDGMENTS

At the top of the list of people to thank is the late and great Bud Youle. Whether he was speaking to me personally or presenting at Red Power events, he left a lasting legacy of colorful, first-person accounts of his experiences with machinery and business. He is the throughline in the book—and much of the color. Godspeed to one of the good ones—it's my hope that his story will inspire people faced with hopeless odds to charge up that hill.

First-person interviews were key to this book's color. Many thanks to those who contributed their stories, including Dr. Glenn Kahle, Kelly Kravig (who gets bonus points for supplying PowerPoint presentations!), Gerry Salzman, Ken Ryan, Cam Beert, Tom Renner, Don Watt, and Steve Warner.

Thanks are due to Lisa Marine and her colleagues at the Wisconsin Historical Society, Emily Orr at the Cooper-Hewitt Smithsonian Design Museum, Sarah Pickett and Adam Summerlott at Case IH, Roger Lewno for sharing his images from the International 50 Series launch, Chuck Pelly, Gregg Montgomery for sharing the fantastic cover art and more, Arthur Carlson at Worcester Polytechnic Institution Archives & Special Collections, Rob Amberg, and the State Historical Society of Iowa, which houses Bill Gillette's work.

Linsey Dodaro did a great job creating a design inspired by the times, and editor Faith Garcia went over the top lavishing editorial care on this book.

Last but far from least, thanks to Joan Hughes for her support of this project and tolerance of having me disappear to write these things.

Credits for Collages

"Battle Royale," pages 8-9

Sketch of John Deere combine, bottom left, *Cooper-Hewitt Smithsonian Design Museum*

Photo of Queen Elizabeth and IH executive, *Case IH*

John Deere tractor with man standing, *Library of Congress*

Blueprints of IH concept tractor, *Keith Burnham*

All others, *Author's Collection*

"The Snoopy Saga," pages 36-37

Various IH advertisements, *Case IH*

Photo of Hinsdale design studio, bottom right, *Adam Summerlott Collection*

Photo of man holding manual, bottom left, *Herbert Starzinger Collection*

All others, *Authors Collection*

"Here Comes the Rain Again", pages 64-65

IH advertisements, *Case IH*

IH model tractor, *Case IH*

John Deere model tractor, *John Westerveld*

All others, *Authors Collection / Clip Art*

"Running Up That Hill," pages 82-83

Men near IH combine, *Roger Lewno Collection*

All others, *Authors Collection*

"The Superdome Showdown," pages 108-9

John Deere tractor in front of brick building, *Cooper-Hewitt Smithsonian Design Museum*

John Deere tractor on circle of tracks, *John Deere*

IH art, ads, and photo of woman with notepad, *Case IH*

Band playing in New Orleans, *Library of Congress*

All others, *Authors Collection*

Front Endpapers (IH machines)

Magnum tractor, center, *Case IH*

Case four-wheel drive, IH two-wheel drive behind it, spacey looking tractor, *Keith Burnham*

All others, *Gregg Montgomery*

Back Endpapers (John Deere machines)

Prototype six-wheel-drive sketch, *Chuck Pelly*

All others, *Cooper-Hewitt Smithsonian Design Museum*

ENDNOTES

Chapter One

1 John "Bud" Youle, phone interview by Lee Klancher and Mary Ham, July 22, 2016.
2 Lee Klancher, "Steiger Experimental Machines: Twin Panther TST650," in *Red 4WD Tractors 1957–2017: High-Horsepower All-Wheel-Drive Tractors from International Harvester, Steiger, and Case IH*, (Austin: Octane Press, 2017), 129–31.
3 "Farm Progress Show," *The Muscatine* (Iowa) *Journal*, September 28, 1977.
4 Don Murray, "Early Research Work, US and France," in *Axial Flow Combine: IH Concept, Development and Production*, (self-pub., 1997), 34.
5 Camiel Beert, video interview by Case IH, Twentieth Anniversary Celebration of the Axial-Flow Combine, 1997.
6 Lee Klancher, "The Red Demon: 1977," in *Red Combines 1915–2020: The Authoritative Guide to International Harvester and Case IH Combines and Harvesting Equipment*, 2nd ed., (Austin: Octane Press, 2021), 163–65.
7 Don Murray, *Axial Flow Combine: IH Concept, Development and Production*, (self-pub., 1997), quoted in Lee Klancher, *Red Combines*, 132.
8 *Red Combines*, 132.
9 Murray, *Axial Flow Combine*, Chapter Eight, 15.
10 Murray, *Axial Flow Combine*, Chapter Eight, 35.
11 Ken Ryan, phone interview by Lee Klancher, February 7, 2013.
12 Ryan, phone interview by Lee Klancher.
13 Don Watt, interview by Lee Klancher at Racine, Illinois, Case IH Facility, May 2014.
14 Ken Ryan, phone interview by Lee Klancher.
15 Lee Klancher, *Farmall Century 1923–2023: The Evolution of Red Tractors and Crawlers in the Golden Age of International Harvester*, (Austin: Octane Press, 2023), 129.
16 Wayne G. Broehl Jr., Appendix: Exhibit 12 and Exhibit 15, in *John Deere's Company – Volume I: From the Steel Plow to the Tractor 1837–1927*, (Austin: Octane Press, 2022) 538, 540.
17 "IHC Annual Report," 1918, housed at Wisconsin Historical Society.
18 "International Harvester's Brooks McCormick: New Spur for a Sluggish Giant," *Businessweek*, March 17, 1975.

Chapter Two

19 "Harvester Sets Profits, Sales Markets in Quarter," *Chicago Tribune*, May 19, 1976. See also "IH Earnings Healthy," *Springfield* (Ohio) *News-Sun*, February 19, 1976.
20 Greg Gackle, "Deere Sales, Earnings Gain," *The Daily Dispatch* (Moline, IL), May 25, 1976.
21 J. P. Kaine, "Four-Wheel Drive Tractor Market Review and Analysis," International Harvester Company 1976 Mid-Year Review, June 1976, housed in the McCormick Archives at the Wisconsin Historical Society.
22 Kelly Kravig, phone interview with Lee Klancher, June 24, 2024.
23 John "Bud" Youle, "2+2 Series," presentation, filmed at Red Power Roundup 2017 in Des Moines, Iowa, June 15, 2017, video, 30:05, https://www.youtube.com/watch?v=pL2Jcbrfnbs.
24 C. W. Gray, *History of Tractors – Farmall: Book II*, (International Harvester), housed at the Wisconsin Historical Society, 15.
25 John J. Bardos and William L. Schubert, "The Design and Development of the International Series 88 Tractors," *SAE Transactions 88* (1979): 2720, https://www.jstor.org/stable/44699085.
26 John "Bud" Youle, phone interview by Lee Klancher and Mary Ham, July 22, 2016.
27 Adam Ambrose, "Steel Firm Sold," *The Times* (Munster, IN), August 3, 1977.
28 "Did IH forgive the McCardell Loan Too Soon," *Chicago Tribune*, November 23, 1980.
29 Glenn Kahle, phone interview by Lee Klancher, March 1, 2013.
30 John "Bud" Youle, "The Big Red Hood of IHC," presentation, filmed at Red Power Roundup 2019 in the Harvester Heritage Tent in Bloomsburg, PA, video, 00:45, https://www.youtube.com/watch?v=rOCZia29IgA.
31 John "Bud" Youle, "Amazing Challenges," presentation, filmed at Red Power Roundup 2015 on the Missouri State Fairgrounds, June 11–12, 2015, video, 01:06, https://www.youtube.com/watch?v=ZAeHwxe878Y.
32 Youle, "The Big Red Hood of IHC," 01:13.
33 Youle, phone interview by Lee Klancher.
34 Youle, "Amazing Challenges," 01:35.
35 Youle, "Amazing Challenges," 07:03.
36 Youle, "Amazing Challenges," 14:12.
37 Youle, "Amazing Challenges," 14:37.
38 Youle, "Amazing Challenges," 29:55.
39 Youle, "Amazing Challenges," 30:07.
40 Youle, "2+2 Series," 22:00.
41 Youle, "Amazing Challenges," 34:35.
42 Youle, "The Big Red Hood of IHC," 38:08.
43 "Profits Trail Deere; Cost Cutting at I. H. to Cost Jobs," *Chicago Tribune*, June 3, 1978.
44 "Harvester Continues Cost-Cutting Crusade," *Chicago Tribune*, December 7, 1978.
45 Youle, "The Big Red Hood of IHC," 42:09.
46 Youle, "Amazing Challenges," 35:37.
47 Youle, "2+2 Series," 25:16.
48 During his presentation at Red Power Roundup 2019, Youle said the cost of the introduction of the 2+2 in January 1978 was $1 million. In his phone interview with Lee Klancher in 2016, he stated the cost for the show was $1.2 million.
49 Youle, "The Big Red Hood of IHC," 43:24.
50 Youle, phone interview by Lee Klancher.
51 Lee Klancher, *Red 4WD Tractors 1957–2017: High-Horsepower All-Wheel-Drive Tractors from International Harvester, Steiger, and Case IH*, (Austin: Octane Press, 2017), 242.
52 Youle, phone interview by Lee Klancher.
53 Youle, "2+2 Series," 56:30.
54 Youle refers to Chuck "Lare" in his presentation. The author was unable to locate anyone who remembers how to spell Chuck's last name, so it is spelled here phonetically.
55 Youle, "Amazing Challenges," 39:45.
56 Youle, "Amazing Challenges," 37:00.
57 Youle, phone interview by Lee Klancher.
58 "IH Annual Report," 1979, housed at Wisconsin Historical Society.
59 Richard G. Hamermesh and Evelyn T. Christiansen, "International Harvester (B2)," *Harvard Business School*, Case 381-054, November 1980 (rev. ed. April 1981).
60 Hamermesh, Appendix: "Archie McCardell's Absolution," 16–18.
61 Barbara Marsh, *A Corporate Tragedy*, (New York: Doubleday, 1985), 205.

Chapter Three

62 "January 4, 1980 Weather History in Washington, D. C.," Weather Spark, https://weatherspark.com/h/d/20957/1980/1/4/Historical-Weather-on-Friday-January-4-1980-in-Washington-D.C.;-United-States#Figures-Temperature.
63 William Robbins, "Grain Farmers Voice Concern," *New York Times*, January 7, 1980.
64 Roger Silkey, interview by Lee Klancher, November 13, 2019.
65 Dale Johnson, interview by Mary Ham, August 18, 2017, quoted in Lee Klancher, *John Deere Evolution: The Design and Engineering of an American Icon* (Austin: Octane Press, 2021), 300–301.
66 R. C. Longworth and Joseph Winski, "Harvester Awaits Return on 'Investment,'" *Chicago Tribune*, April 20, 1980.
67 Lee Klancher, *Red Tractors 1958–2022: The Authoritative Guide to International Harvester and Case IH Tractors*, 3rd ed., (Austin: Octane Press, 2022), 21, 248.
68 Steve Warner, phone interview by Lee Klancher, January 22, 2013.
69 "It's Official: The Recession is Really Here," *Chicago Tribune*, June 8, 1980.

70 "3788 Tractor," International Harvester memo, June 9, 1980, housed in the McCormick Archives at the Wisconsin Historical Society.
71 Winston Williams, "It's Been Rough Going for the Big-Rig Makers," *The Press Democrat* (Santa Rosa, CA), October 12,1980.
72 Mark Potts, "Harvester Will Close 2 Plants, Idling 1,700," *Chicago Tribune*, September 22, 1980.
73 Joseph Winski, "Harvester Retires Its Last McCormick," *Chicago Tribune*, November 17, 1980.
74 Joseph Winski, "Did I-H Forgive the McCardell Loan Too Soon?," *Chicago Tribune*, November 23, 1980.
75 Winski, "Did I-H Forgive."
76 Kate Corcoran, "1980: Tough on Balance Sheets," *The Daily Dispatch* (Moline, IL), December 31,1980.
77 Winski, "Did I-H Forgive."

Chapter Four

78 "International Harvester Looks for Better Days," *Wichita* (Kansas) *Eagle-Beacon*, January 1, 1981.
79 Winston Williams, "Harvester Salaries Cut By 20%," *New York Times*, January 15, 1981.
80 "Loan Forgiven by Harvester," *New York Times*, January 23, 1981.
81 "Loan Forgiven by Harvester," *New York Times*.
82 James O'Shea, "Moody's Lowers Debt Rating on Harvester Co.," *Chicago Tribune*, February 4, 1981.
83 John "Bud" Youle, "The Big Red Hood of IHC," presentation, filmed at Red Power Roundup 2019 in the Harvester Heritage Tent in Bloomsburg, PA, video, 36:00, https://www.youtube.com/watch?v=rOCZia29IgA.
84 Youle, "The Big Red Hood of IHC," 32:40.
85 "How Many Mexico Tractors Are Around," Red Power Forum, 2014, https://www.redpowermagazine.com/forums/topic/84704-how-many-mexico-tractors-are-around/; "A Taco tractor?," Red Power Forum, 2008, https://www.redpowermagazine.com/forums/topic/37225-a-taco-tractor/; "Mexican 86 Series IH," AgTalk, 2013, https://talk.newagtalk.com/forums/thread-view.asp?tid=436663&DisplayType=flat; Gerry Salzman, phone interview by Lee Klancher, May 2, 2024.
86 Youle, "The Big Red Hood of IHC," 33:00.
87 "IH Annual Report," 1981, housed in the McCormick Archives at the Wisconsin Historical Society.
88 Winston Williams, "Caterpillar to Acquire Harvester's Solar Unit," *New York Times*, May 15, 1981.
89 Joseph Winski and Pam Sebastian, "Harvester Omits Dividend, Posts $96.4 Million Loss," *Chicago Tribune*, February 20, 1981.
90 The 1981 IH Annual Report did not list research and development expenses, as was done on most previous annual reports.
91 Wayne G. Broehl Jr., Appendix: Exhibit 36, "Research and Development Expenses as Percent of Sales and Dollars Per Employee 1977–1981," in *John Deere's Company – Volume II: From Johnny Popper to the Iron Horses 1928–1982*, (Austin: Octane Press, 2023), 346.
92 Lee Klancher, *Red Tractors 1958–2022: The Authoritative Guide to International Harvester and Case IH Tractors*, 3rd ed., (Austin: Octane Press, 2022), 203–26.
93 Oscar H. Will III, "50 Series Development," in *Red Tractors 1958–2022: The Authoritative Guide to International Harvester and Case IH Tractors*, 3rd ed., (Austin: Octane Press, 2022), 203–7.
94 Glenn Kahle, phone interview by Lee Klancher, February 15, 2013.
95 Kahle, phone interview by Lee Klancher, February 15, 2013.
96 "Q&A with Mike Schachtrup," *Peoria Magazine*, April 2018.
97 Youle, "The Big Red Hood of IHC," 39:00.
98 Youle, "The Big Red Hood of IHC," 39:06.
99 John "Bud" Youle, "Amazing Challenges," presentation, filmed at Red Power Roundup 2015 on the Missouri State Fairgrounds, June 11–12, 2015, video, 43:01. https://www.youtube.com/watch?v=ZAeHwxe878Y.
100 In Youle's stories, it's always the CEO, governor, or president on the phone.
101 Youle, "The Big Red Hood of IHC," 52:34.
102 Youle, "The Big Red Hood of IHC," 53:40.
103 Youle, "Amazing Challenges," 54:45.
104 Youle, "Amazing Challenges," 45:26.
105 Salzman, phone interview by Lee Klancher.
106 Pam Sebastian, "Smaller Harvester Has Better Chance to Survive: Analysts," *Chicago Tribune*, December 27, 1981.
107 Salzman, phone interview by Lee Klancher.
108 Kahle, phone interview by Lee Klancher, February 15, 2013.
109 Pam Sebastian, "Harvester Reaps Bountifully from Retirees' Return," *Chicago Tribune*, June 27, 1983.

Chapter Five

110 "UAW Lists Issues for Harvester," *The Courier-Journal* (Louisville, KY), January 14, 1982.
111 James L. Rowe Jr., "Plowing Up the Seeds of Poverty," *Washington Post*, January 3, 1982.
112 James L. Rowe Jr., "McCardell Resigns as Chairman of Harvester," *Washington Post*, May 3, 1982.
113 Pam Sebastian, "McCardell Heeds Harvester Request to Resign," *Chicago Tribune*, May 4, 1982.
114 Sebastian, "McCardell Heeds."
115 David Young, "New IH Chief 'Intuitive Guy,'" *Chicago Tribune*, May 4, 1982.
116 Mark Potts, "McCardell Gone, but IH Needs Intensive Care," *Chicago Tribune*, May 9, 1982.
117 Gerry Salzman, phone interview by Lee Klancher, May 2, 2024.
118 Robert Gannon, "Solving the Puzzle of El Nino," *Popular Science*, September 1986.
119 "Harvester to Sell 5 Parts Centers," *New York Times*, June 25, 1982.
120 John "Bud" Youle, "The Big Red Hood of IHC," presentation, filmed at Red Power Roundup 2019 in the Harvester Heritage Tent in Bloomsburg, PA, video, 01:01, https://www.youtube.com/watch?v=rOCZia29IgA.
121 Youle, "The Big Red Hood of IHC," 01:03.
122 John "Bud" Youle, phone interview by Lee Klancher and Mary Ham, July 22, 2016.
123 Dale Johnson, phone interview by Mary Ham, August 18, 2017.
124 Youle, phone interview by Lee Klancher.
125 Tom Renner, phone interview by Lee Klancher, May 14, 2024.
126 "John Deere 4850 MFWD Opinions," AgTalk, March 12, 2016, https://talk.newagtalk.com/forums/thread-view.asp?tid=618468&DisplayType=flat&setcookie=1.
127 Johnson, phone interview by Mary Ham.
128 Youle, phone interview by Lee Klancher.
129 Youle, phone interview by Lee Klancher.
130 Youle, "The Big Red Hood of IHC," 01:05.
131 John "Bud" Youle, "Amazing Challenges," presentation, filmed at Red Power Roundup 2015 on the Missouri State Fairgrounds, June 11–12, 2015, video, 59:00. https://www.youtube.com/watch?v=ZAeHwxe878Y.
132 Youle, phone interview by Lee Klancher.
133 Youle, phone interview by Lee Klancher.
134 Salzman, phone interview by Lee Klancher.

Epilogue

135 "Q&A with Mike Schachtrup," *Peoria Magazine*, April 2018.
136 John "Bud" Youle, "Amazing Challenges," presentation, filmed at Red Power Roundup 2015 on the Missouri State Fairgrounds, June 11–12, 2015, video, 01:20. https://www.youtube.com/watch?v=ZAeHwxe878Y.
137 Tim Moran, "Deere & Co. Hopes to Harvest New Tractor Sales," *Rock Island Argus* (Moline, IL), August 1, 1982.
138 John "Bud" Youle, "The Big Red Hood of IHC," presentation, filmed at Red Power Roundup 2019 in the Harvester Heritage Tent in Bloomsburg, PA, video, 01:16, https://www.youtube.com/watch?v=rOCZia29IgA.

INDEX

JOHN DEERE
50-X